A Building Craft Foundation

Peter Brett

Nelson Thornes
a Wolters Kluwer business

First published in 1991 by:
Stanley Thornes (Publishers) Ltd
Second edition 2002

Third edition published in 2007 by:
Nelson Thornes Ltd
Delta Place
27 Bath Road
CHELTENHAM
GL53 7TH
United Kingdom

07 08 09 10 11 / 10 9 8 7 6 5 4 3 2 1

A catalogue record for this book is available from the British Library

ISBN 978 0 7487 8184 3

Cover photos (from top left row by row): Corbis CI 183 (NT) 2, 4, 6 ; Brand X HC 274
(NT) 1, 3, 5, 9; Ingram ILS V1 CD2 CON (NT) 8; Ingram ILR V1 CD2 BS (NT); Wood
texture along top Corel (NT) 7.

New illustrations by Angela Lumley
Archive illustrations include artwork by Peters and Zabransky (UK) Ltd and Richard
Morris, with colour added by Pantek Arts Ltd.

Page make-up by Pantek Arts Ltd, Maidstone, Kent

Printed and bound in Slovenia by Korotan

Picture credits
George Disario/Corbis, p. vii
The £1 and 10 pence coin on page 204 are reproduced courtesy of The Royal Mint;
the £5 note on the same page is reproduced courtesy of the Bank of England.
Alamy, p. 72.

Contents

Acknowledgements

My sincere thanks go to: my wife Christine for her assistance, support and constant encouragement; my daughter Sarah; grandchildren Matthew, Chris and Rebecca and son James and his partner Claire for their support, patience and sanity checks; my colleagues and associates past and present for their continued support of my work and motivation to continue.

Finally 'all the best for the future' to all who use this book, I trust it provides you with some of the help and motivation required to succeed in the construction industry.

Peter Brett

Introduction

National Vocational Qualifications (NVQs) in Construction

These qualifications focus on practical skills and knowledge. They have been developed and approved by people who work in the construction industry.

Construction NVQs are available in England, Wales and Northern Ireland. Scotland has SVQs, which work in a similar way.

There are three levels of NVQs for construction crafts and operatives:

◆ Level 1 is seen as a 'foundation' to the construction industry, consisting of common core skills and occupational basic skills.
◆ Level 2 consists of common core skills and units of competence in a recognizable work role.
◆ Level 3 consists of further common core skills, plus a more complex set of units of competence in a recognizable work role, including some work of a supervisory nature.

Awarding body

CITB-Construction Skills and City & Guilds are the joint awarding body for the construction industry. CITB-Construction Skills are also responsible for the setting of standards for Craft and Operative NVQs.

Work roles

Each construction NVQ focuses on an individual work role, for example:

◆ Bricklaying
◆ Site Carpentry
◆ Bench Joinery
◆ Painting and Decorating
◆ Plastering
◆ Shop Fitting etc.

Construction NVQ make up

Each work role is made up of a number of individual **Units of Competence**, for example:

All Mandatory Units must be undertaken plus a number of the options (three in the case of site carpentry). The number of option units in a work role and the number that are required to be undertaken will vary depending on the extent of the particular work role.

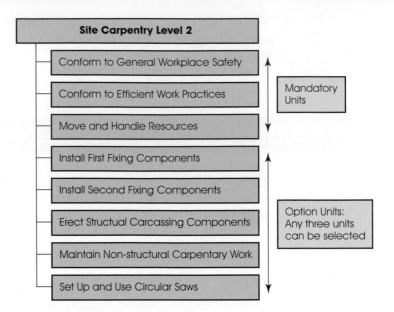

Figure 0.1

Unit of competence make up

In order to set out exactly what is contained in a unit and also make it easier to assess, each unit begins with a **description**, for example:

Conform to General Workplace Safety

This unit is about:

◆ awareness of relevant current statutory requirements and official guidance
◆ personal responsibilities relating to workplace safety, wearing appropriate personal protective equipment (PPE) and compliance with warning/safety signs
◆ personal behaviour in the workplace
◆ security in the workplace

The description is followed by a number of statements:

Performance criteria – these state exactly what you must be able to do.

Identify hazards

Scope of performance – this sets out what evidence is required to meet each of the performance criteria. The majority of this evidence must be from the workplace; simulation evidence is only allowed in limited circumstances.

Hazards, associated with the workplace and occupations at work, are recorded and/or reported

Knowledge and understanding relating to performance criteria – this links in general terms the knowledge and understanding required to back up the performance criteria.

You must know and understand:

◆ the **hazards** associated with the occupational area
◆ the method of **reporting** hazards in the workplace

Scope of knowledge and understanding – this uses the key words contained in the knowledge and understanding statements (shown in **bold type**) and expands them to cover the scope of what is expected of a competent worker in the construction industry.

Hazards:

◆ Associated with resources, workplace, environment, substances, equipment, obstructions, storage, services and work activities.

Reporting

◆ Organisational reporting procedures and statutory requirements.

Collecting evidence

You will need to collect evidence from your workplace of your satisfactory performance in each performance criteria of a unit of competence. This should be inserted in a portfolio and referenced to each unit of competence. Evidence must confirm that your practical skills meet the appropriate performance criteria. Simulation evidence in a training environment is only allowed in a limited range of topics.

Evidence can come from any of the following people:

◆ employers
◆ managers
◆ supervisors
◆ skilled work colleagues
◆ work-based recorder
◆ client.

Figure 0.2

Suitable types of evidence

You should include in your portfolio as much evidence as possible, from more than one of the following, for each performance criteria:

◆ **Time Sheets** – detailing the work you have undertaken; for these to be valid they must be signed by you and the work-based recorder.
◆ **Drawings** – of the work you have undertaken; these should be supported by a witness testimony.
◆ **Photographs** – of the work you have undertaken; ideally with you in the photograph. To be valid, photographs should be supported by a statement, containing a brief description of the work, details of where and when you carried it out and be signed by you and either the work-based recorder, manager, supervisor, skilled worker or the client.

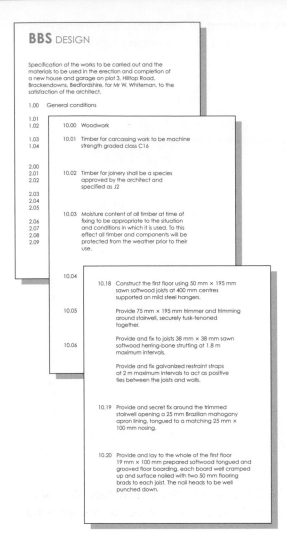

Figure 0.3
Extracts from a specification

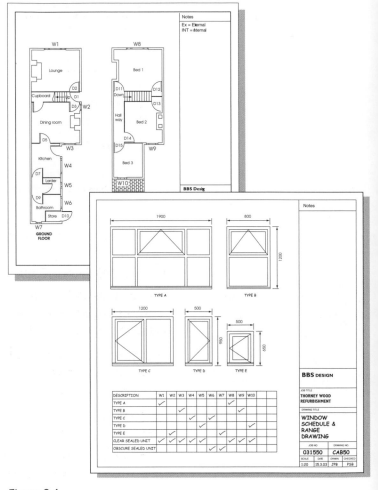

Figure 0.4
Building schedule

◆ **Associated Documentation** – used or produced as part of the work you have undertaken, such as specifications, forms and reports completed.

◆ **Witness Testimony** – a statement by a responsible person confirming that you have undertaken certain work activities; these should include, wherever possible, a detailed description of the work you carried out.

Where an assessor considers your evidence as insufficient in either quality or quantity, you may be asked to undertake simulated activities in order to demonstrate/reinforce your competence in particular performance criteria.

Introduction

T. Joycee Construction

Ridge House
Norton Road
Cheltenham
GL59 1DB

To whom it may concern:

I can confirm, that between 15 March and 8 November 2006 James Oakley worked on the refurbishment contract at the Rivermead Estate.

James was involved in the replacement of casement windows and internal window boards. He carried this work out to a competent standard at all times.

This work was undertaken in occupied houses, feedback from the tenants concerning James's communication with them and his consideration shown to their property, including the cleanliness of his work was always exemplary.

In addition James assisted me in the general day-to-day organization of the working environment, including the scheduling of the work and the safety induction of new staff. Indeed he always set a fine example by wearing at all times his safety helmet, boots and high visibility vest.

Yours Faithfully

Chris Heath

Chris Heath
(Site Project Manager)

Figure 0.5
Witness testimony

The assessment process

The joint awarding body CITB-Construction Skills and City & Guilds approves organizations to carry out assessment of people for an NVQ award in construction. Typically these are:

- further education colleges
- private training providers
- construction companies.

Once approved, these are known as Assessment Organizations. Their assessment work will involve the following personnel:

- *Assessors* – these are people who are occupationally competent in the work role in which you are being assessed and also qualified in the assessment process. Their role is to decide whether you are competent in each performance criterion. They will also observe you in the workplace to ensure you are carrying out the full range of activities to create the required evidence portfolio.
- *Internal verifier* – this is the person who is responsible in an assessment organization for ensuring the quality of the assessments carried out by the assessors.
- *Work-based recorders* – these are people in the workplace whose employer has given them the responsibility of authenticating the evidence that a candidate is collecting for their portfolio.
- *External verifiers* – they are employed by the joint awarding body to monitor the whole assessment process and ensure that each assessment organization is working to the standards set.

How to use this book

This book covers the three mandatory core units, which are common to all work roles as well as an introduction to the Construction Industry and Numerical Skills; separate books are available for the Wood Occupations at Level 1 and both Site Carpentry and Bench Joinery at Level 2.

These books are intended to be supported by:

◆ classroom activities
◆ tutor reinforcement and guidance
◆ group discussion
◆ films, slides and videos
◆ text books
◆ independent study/research
◆ practical activities.

You will be working towards one or more units at a time as required. Discuss its content with your group, tutor and/or friends wherever possible. Attempt to answer the learning activities for that unit. Progressively work through all the units, discussing them and answering the assessment activities as you go. At the same time, you should be working on the matching practical activities in the workplace and collecting the required evidence.

This process is intended to aid learning and enable you to evaluate your understanding of the particular topic and to check your progress through the units. Where you are unable to answer a question, further reading and discussion of the topic is required.

Independent study/research

Browsing the Internet via a computer is an excellent means of accessing other sources of information as part of your research: simply type in the website address of the company or organization into a web browser and you will be connected to their website.

Try some of the following sites:

◆ Building Regulations: www.planningportal.gov.uk
◆ British Standards: www.bsi-global.com
◆ Building Research Establishment: www.bre.co.uk
◆ Construction training and careers: www.citb-constructionskills.co.uk and www.city-guilds.co.uk
◆ Government publications: www.tso.co.uk
◆ Health and safety: www.hse.gov.uk
◆ Building materials and components: www.buildingcentre.co.uk
◆ Employment Rights and Trade Unions www.worksmart.org.uk

If you don't know the exact website address of the organization you are looking for, or you simply wish to find out more information on a subject, you could use a search engine to find the web pages. Search engines use key words to find information on a subject. Enter a key word or words such as 'doors', 'windows', 'stairs' or 'strength grading' or 'timber', etc. or the name of a company/organization, and it searches the Internet for information about your key words or name. You are then presented with a list of relevant websites that you can click on, which link you to the appropriate information pages.

Types of learning activity

The learning activities used in this book should be completed on loose-leaf paper and included as part of your portfolio of evidence. They are divided into the following:

◆ Measuring up. Questions at the end of a major topic or units, which enable you to evaluate your understanding of a recently completed topic and to check your progress through the units. 'Measuring-up' questions are either multiple-choice questions or short answer questions.

◆ Activity. An extended learning task normally at the end of a unit, which has been designed to reinforce your technical and communication skills in day-to-day work situations.

Multiple-choice questions normally consist of a statement or question followed by four possible answers. Only one answer is correct; the others are distracters. You have to select the most appropriate letter as your response.

Example 1:

The shape of a *Prohibition* safety sign is:

a) Square
b) Round
c) Triangular
d) Rectangular.

As Round is the correct answer your response should be

Example 2:

The safety sign illustrated in the figure indicates:

a) Smoking permitted in this area
b) Smoking prohibited in this area
c) Smoking is bad for your health
d) Smoking is not recommended.

Figure 0.6

The correct response is (b).

Occasionally variants on the four-option multiple-choice question are used, as in the following examples.

How to use this book

How to use this book

Example 3:

Match the items in **list one** with the items in **list two**.

List one *refer to illustrations*;

W   X

Y   Z

Figure 0.7

List two

1. Irritant

2. Corrosive

3. Toxic

4. Explosive

5. Highly Flammable.

The correct match is:

	W	X	Y	Z
a)	5	4	3	2
b)	4	5	1	3
c)	5	4	1	3
d)	4	5	3	2

This question requires you to work through the lists matching the items up (it is usual for the lists to be of different lengths). In this example:

W is 4
X is 5
Y is 3
Z is 2

Therefore the correct response is (d).

Example 4:

Statement: The use of PPE should be regarded as the last resort in controlling risk of personal injury.

Reason: The first consideration is to undertake a risk assessment to eliminate or control risks.

a)	Statement true	Reason true
b)	Statement false	Reason false
c)	Statement true	Reason false
d)	Statement false	Reason true

This type of question comprises a statement followed by a reason, where both the statement and reason can be true or false. You are required to select the appropriate response.

In this example both the statement and reason are true, therefore the correct response is (a).

measuring up

Short-answer questions consist of a task to which a short written answer is required. The length will vary depending on the 'doing' word in the task:

Name or **List** normally requires one or two words for each item;

State, **Define**, **Describe** or **Explain** will require a short sentence or two;

In addition, sketches can be added to any written answer to aid clarification.

Draw or **Sketch** will require you to produce an illustration.

Example 5:

Name the person who designs and supervises the construction of a building for the client.

Typical answer: The Architect.

Example 6:

Define the term 'the building team'.

Typical answer: The team of professionals who work together to produce the required building or structure. Normally consists of the following parties: client, architect, quantity surveyor, specialist engineers, clerk of works, local authority health and safety inspector, building contractor, subcontractors and material suppliers.

Example 7:

Make a sketch to show the difference in size between a brick and a block.

Typical answer:

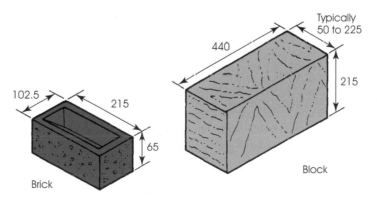

Figure 0.8

These *activities* are a combination of short answer questions on the same topic. They normally commence with a statement containing a certain amount of relevant information, which is designed to set the scene for the question. This is then followed by a series of related sub questions in logical order. The length of the expected answer to each subpart will again vary, depending on the topic and the wording of the question, from one or two words to a paragraph, a form to complete, a sketch, a calculation, or a combination of any of these. At each stage, the wording of the question will make it clear what is required. Blank forms for completion and inclusion in your portfolio of evidence may be downloaded from www.nelsonthornes.com/carpentry.

Example 8:

The general location drawing and programme chart for a house, which is under construction are illustrated in the following figures.

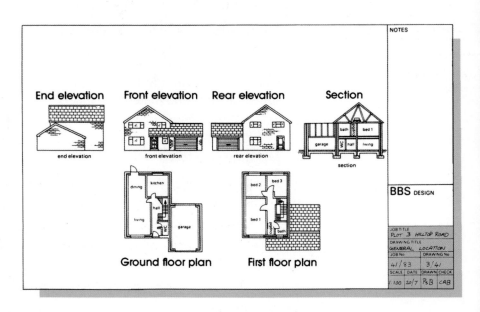

Figure 0.9
General location drawing

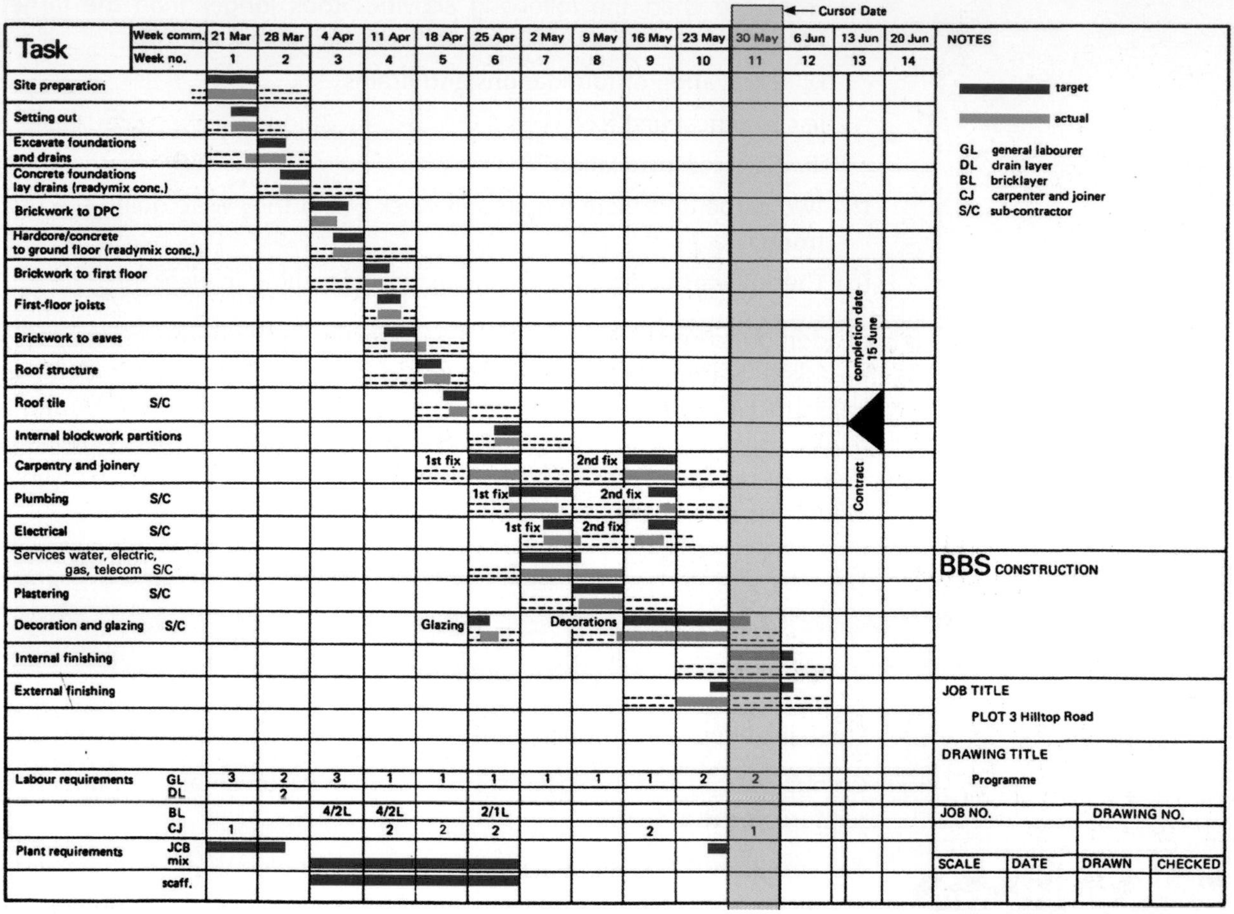

Figure 0.10
Programme chart

Task	Week comm.	21 Mar	28 Mar	4 Apr	11 Apr	18 Apr	25 Apr	2 May	9 May	16 May	23 May	30 May	6 Jun	13 Jun	20 Jun	NOTES
	Week no.	1	2	3	4	5	6	7	8	9	10	11	12	13	14	
Site preparation																target / actual
Setting out																GL general labourer
Excavate foundations and drains																DL drain layer
Concrete foundations lay drains (readymix conc.)																BL bricklayer
Brickwork to DPC																CJ carpenter and joiner
Hardcore/concrete to ground floor (readymix conc.)																S/C sub-contractor
Brickwork to first floor																
First-floor joists																
Brickwork to eaves																
Roof structure																
Roof tile S/C																
Internal blockwork partitions																
Carpentry and joinery						1st fix			2nd fix							
Plumbing S/C							1st fix			2nd fix						
Electrical S/C								1st fix	2nd fix							
Services water, electric, gas, telecom S/C																BBS CONSTRUCTION
Plastering S/C																
Decoration and glazing S/C						Glazing		Decorations								
Internal finishing																
External finishing																JOB TITLE: PLOT 3 Hilltop Road

Cursor Date — week 11 (30 May). completion date 15 June. Contract.

Labour requirements		1	2	3	4	5	6	7	8	9	10	11	12	13	14
GL		3	2	3	1	1	1	1	1	1	2	2			
DL			2												
BL				4/2L	4/2L		2/1L								
CJ		1			2	2	2			2		1			
Plant requirements															
JCB mix															
scaff.															

DRAWING TITLE: Programme
JOB NO.   DRAWING NO.
SCALE | DATE | DRAWN | CHECKED

(a) Describe the type of construction and style of building.

(b) Determine the number of internal and external doors required excluding the garage.

(c) Name the type of chart used for the programme.

(d) What is the planned contract completion week number and date?

(e) What activities took longer than the target time?

(f) State how many carpenters were required in week 6 and what activity they were engaged in.

(g) Name the activity that the JCB was being used for in week 10.

Typical answer:

(a) The building style is a detached two-storey house. It is built using a cavity wall cellular construction, which is also known as solid or masswall construction.

(b) There are two external doors and eight internal doors.

(c) The programme is shown on a Bar or Gantt chart.

(d) The contract is planned to be complete in week thirteen on the 15th of June.

(e) From the chart the following activities took longer than the target time:

    i. Excavation of foundations and drains

    ii. Electrical first fix

    iii. Services installation

(f) Two carpenters were required in week 6 and they were doing the first fixing

(g) The JCB was being used for the external finishing in week 10.

In addition to completing the learning activities in the book, you may also be asked oral questions by an approved assessor. This is often done to gain further evidence of your written response or can be asked during a review of your portfolio to gain supplementary evidence: these questions normally take the form of: 'How did you ... ?'; 'Why did you ... ?; 'What would you do in the following circumstances ... ?', etc.

Other learning features used in this book

These include the following:

Colour enhanced **illustrations** and **documents** as an aid to clarity and reinforcement of text.

did you know? boxes in the margin, which define new words or highlight key facts.

safety tip boxes in the margin, highlighting facts for you to follow or be aware of when undertaking practical tasks.

Example

Worked examples included in the text for use as a guide when answering questions or undertaking other tasks.

Always read safety method statements and take appropriate action.

Hazards are something with the potential to cause harm.

Harm can vary in its severity; some hazards can cause death, others illness or disability or maybe only cuts or bruises.

Risk is concerned with the severity of harm and the likelihood of it happening.

example

A 3.114 m length of wall is to be panelled with 95 mm (90 mm covering width) matchboard. Divide the length of wall in millimetres by the covering width of the board.

$$3114 \div 90 = 34.6$$

Therefore 35 boards are required $= $ 33 whole boards and two end boards.

$$= 1.6 \times 90 \div 2$$

Width of cut ends boards $=$ 72 mm

The Construction Industry

This chapter is intended to provide the new entrant with an overview of the construction industry. Although its content is not assessed directly, knowledge of its contents is assumed in other assessed units. It is concerned with the personnel, principles and terms that are encountered on a day-to-day basis.

In this chapter you will cover the following range of topics:

◆ Introduction to the construction industry
◆ Construction occupations
◆ The building team
◆ Principles of construction.

Introduction to the construction industry

Government statistics divide the construction or 'building' industry into five main areas, by the type of firm or organization undertaking the work. They all work to the same purpose, which is to provide and maintain accommodation and other services to meet the varying needs of the population as a whole.

◆ **Construction and repair of buildings** – the construction, improvement and repair of both residential and non-residential buildings, including the specialist organizations of bricklaying, carpentry, general building maintenance, roofing, scaffolding and the erection of framed structures for buildings.

◆ **Civil engineering** – the construction and maintenance of roads, car parks, bridges, railways, airport runways, and works associated with dams, reservoirs, harbours, rivers, canals, irrigation and land drainage. In addition the laying of pipelines or cables for sewers, gas and water mains and electricity, including overhead lines and supporting structures.

◆ **Installation of fixtures and fittings** – the installation of fixtures and fittings for electricity, gas, plumbing, heating and ventilation, sound and thermal insulation.

◆ **Building completion work** – firms specializing in work having a completion role to a building, including on-site carpentry and joinery, painting and decorating, glazing, plastering, tiling and flooring, etc.

did you know?

New – a building that has just been built.

Maintenance – the repairs undertaken to a building and its services to keep it at an acceptable standard so that it may fulfil its function.

Refurbishment – to bring an existing building up to standard, or make it suitable for a new use by **renovation**.

Restoration – to bring an existing building back to its original condition.

Conservation – the preservation of the environment and management of natural resources.

◆ *General construction and demolition work* – a general classification to include firms and other organizations engaged in building and civil engineering whose work is not sufficiently specialized to be classified in one of the previous four. Demolition work and direct labour establishments of local authorities and government departments are also included under this heading.

The value of this work in the UK is about £85 billion per year; at the time of writing. About 54% of the annual total is for new work and 46% for repairs and maintenance. The pie chart gives a further percentage breakdown of these statistics.

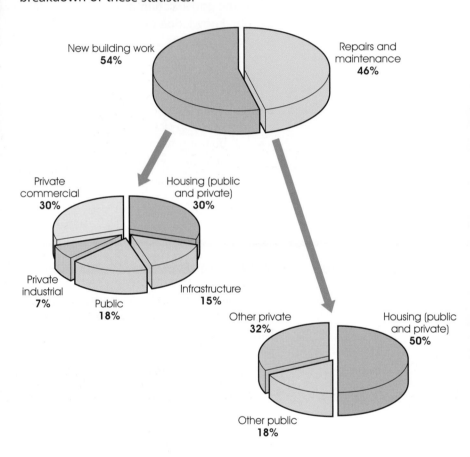

Figure 1.1

The work of the construction industry can also be divided into work carried out either for the public or private sectors.

◆ *Public work* – for any public authority such as local and central government departments, public utilities, nationalized industries, universities, new town corporations and housing associations, etc.

◆ *Private work* – for a private owner, organization or developer and includes all work undertaken by businesses on their own initiative. In addition work carried out under the government's PFI (Private Financial Initiative) is included. PFI is work where an individual firm or consortium takes on the responsibility for providing a public service, typical examples being in health care and education where the private sector is actively engaged in building, equipping and maintaining the infrastructure.

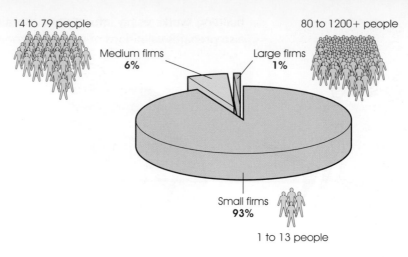

14 to 79 people

80 to 1200+ people

Medium firms
6%

Large firms
1%

Small firms
93%

1 to 13 people

Figure 1.2

Classification of construction companies by size

At the time of writing there are over 166,000 registered firms in the UK of varying size undertaking building work. The vast majority, 93%, are small firms that employ 1 to 13 people; 6% are medium size firms employing 14 to 79 people. Large companies employing 80 or more people account for the remaining 1%.

Many of the smaller firms are self-employed people or small organizations, which are employed by the medium and larger firms on a subcontract basis.

In total the building industry employs almost two million workers, representing 7% of the country's working population. Of these, 60% are directly employed and 40% self-employed. The top occupations are carpenters and joiners with over 10% of the total; plumbers 7.5%; electricians 7%; managers 6.5%; painters and decorators 6%; and bricklayers 5%.

Construction occupations

The construction industry offers employment in four distinct career areas: professional, technician, building crafts, building operatives.

Professional

These graduate entry positions include the following:

- ◆ *Architect* – designs and supervises the construction of buildings.
- ◆ *Engineer* – can be a civil engineer (concerned with roads and railways, etc.), a structural engineer (concerned with the structural aspects of a building's design) or a service engineer who plans building service systems.
- ◆ *Surveyor* – can be a land surveyor (who determines positions for buildings, roads and bridges, etc.), a building surveyor (who is concerned with the administration of maintenance and adaptation works as well as new buildings) or a quantity surveyor (who measures and describes

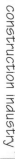

The construction industry

Chapter 1

building works using information contained on architects' drawings; also prepares valuations of works in progress).

Figure 1.3

Figure 1.4

Figure 1.5

Technician

This is the link level in the industry between the professional and craft areas. The main job functions of technicians are as follows:

- *Architectural technician* – involved with the interpretation and presentation of the architect's design information, into a form suitable for use by the builder.

- *Building technician* – involved with the estimating, purchasing, site surveying, site management and documentation of building works.

- *Building surveying technician* – may specialize in building maintenance, building control or structural surveys, etc.

- *Quantity surveying technician* – calculates costs and payments for building works.

Building crafts

Figure 1.6

The building crafts involve the skilled operatives who work with specific materials and actually undertake the physical tasks of constructing a building. The main examples are as follows:

- *Bricklayer* – works with bricks and mortar to construct all types of walling; also concerned with maintenance and adaptation of existing works.

- *Carpenter and/or joiner* – works with timber, other allied materials, metal/plastic items and ironmongery. Makes, fixes and repairs all timber components in buildings. Carpenters work on building sites, whereas joiners work mainly in a workshop at the bench.

Figure 1.7

- *Electrician* – works with metals, plastics, wire and cables, and installs and maintains electrical systems.

- *Formworker* – works with timber, metal and plastic, etc. to produce a structure that supports and shapes wet concrete until it has become self-supporting.

Figure 1.8

Figure 1.9

Figure 1.10

◆ *Painter and decorator* – works with paint, paper, fabrics and fillers, to decorate or redecorate new and existing works; sometimes glazes windows and carries out sign writing.

◆ *Plasterer* – works with plaster, cement mixes, plasterboard and expanded metal, to finish walls, ceilings and floors; also makes and fixes plaster decorations.

◆ *Plumber* – works with metals, plastics and ceramics; installs tanks, baths, toilets, sinks, basins, rainwater goods, boilers, radiators and gas appliances; also cuts and fixes sheet-metal roof covering and flashing and sometimes glazing; also maintains existing works.

◆ *Roof slater/tiler* – works with felt, timber, metals, mortar and a wide variety of slates and tiles; covers new or existing pitched roofs with slates or tiles; also maintains existing works.

◆ *Shopfitter* – works with timber, metal, glass and plastics, etc.; makes and installs shop fronts and interiors, also for banks, hotels, offices and restaurants.

◆ *Stonemason* – works with stone and mortar; a 'banker' cuts and smoothes stone while a 'fixer' erects prepared stones.

Figure 1.11

Figure 1.12

Figure 1.13

The construction industry | Chapter 1

Figure 1.14

Figure 1.15

Figure 1.16

◆ *Woodworking machinist* – operates a wide range of woodworking machines; prepares timber for the production of timber-building components.

◆ *Building operatives* – there are two main types employed on-site. The **general building operative** uses various items of plant, e.g. hand tools, power tools, compressors and concreting equipment, etc.; mixes concrete, mortar and plaster; lays drainage, kerb stones and concrete, etc.; offloads materials and transports around site; also generally assists the work of craft operatives. The **specialist building operative** carries out specialist building operations, e.g. ceiling fixer, dry liner, glazier, mastic asphalter, built-up felt roofer, plant mechanic, roof sheeter and cladder, scaffolder, wall and floor tiler, etc.

measuring up

1. Name **THREE** of the main areas of work that make up the building industry.

2. State the difference between private and public sector building work.

3. Name **THREE** examples of building work.

4. Name **THREE** examples of civil engineering work.

5. State the difference between maintenance, refurbishment and restoration.

6. State the main purpose of the construction industry.

7. State the classification of a contractor with 100 employees.

The building team

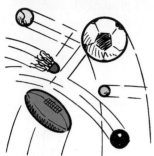

Figure 1.17

Planning coordinator (CDM coordinator) – is the person appointed by the client to prepare a safety plan and file and coordinate with the principle contractor on health and safety matters, in accordance with the Construction, Design and Management Regulations.

Principle contractor – the main contractor who has been awarded a contract to undertake building work, they may appoint other contractors or subcontractors to do part of the work. The principle contractor may ensure cooperation on health and safety matters between all persons on-site.

Bill of quantities – a document prepared by the quantity surveyor, it gives a description and measure of quantities of labour, materials and other items required to carry out a building contract. It is based on the drawings, specifications and schedules and forms part of the contract documents.

The construction of a building is a complex process, which requires a team of professionals working together to produce the desired results. Collectively they are known as the building team and are a combination of the following parties:

Client

This is the building owner; the person or persons who have an actual need for building work, e.g. the construction of a new house, office block, factory, etc. or extensions, repairs and alterations to existing buildings. The client is the most important member of the building team; without the client the work would not exist. The client is responsible for commissioning and the overall financing of the work and in effect employs either directly or indirectly the entire team. In particular they have the specific responsibility under the Construction (Design and Management) Regulations for appointing the **planning coordinator** and nominating the **principal contractor**.

Architect

The architect is the client's agent and is considered to be the leader of the building team. The role of an architect is to interpret the client's requirements, translate them along with other specialist designers into a building form and generally supervise all aspects of the work until it is completed. In addition architects or a member of their team are often appointed by the client under the Construction (Design and Management) Regulations as the planning coordinator, responsible for drawing up a safety plan, coordinating with the principal contractor regarding on-site **health and safety** and the compilation of a health and safety file.

Quantity surveyor

In effect, the quantity surveyor (QS) is the client's economic consultant or accountant. This specialist surveyor advises during the design stage as to how the building may be constructed within the client's budget, and measures the quantity of labour and materials necessary to complete the building work from drawings and other information prepared and supplied by the architect. These quantities are incorporated into a document known as the **bill of quantities**, which is used by building contractors when pricing the building work. During the contract, the quantity surveyor will measure and prepare valuations of the work carried out to date to enable interim payments to be made to the building contractor and at the end of the building contract they will prepare the final account for presentation to the client. In addition, the quantity surveyor will advise the architect on the cost of any additional work or variations.

The construction industry

Chapter 1

did you know?

There are three areas of building control:

Planning controls – restrict the type, position and use of buildings or development in relation to the environment.

Building regulations – set out functional requirements for the design and construction of buildings to ensure the health and safety of people in and around them, promote energy efficiency and contribute to meetings the needs of disabled people.

Health and safety controls – are concerned with the health and safety of all perons at their place of work and protecting other people, such as visitors and the general public, from risks occurring through work activities.

Figure 1.18

Consulting (specialist) engineers

These are engaged as part of the design team to assist the architect in the design of the building within their specialist fields, e.g. civil engineers, structural engineers and service engineers. They will prepare drawings and calculations to enable specialist contractors to quote for these areas of work. In addition, during the contract the specialist engineers will make regular inspections to ensure the installation is carried out in accordance with the design.

Clerk of works (COW)

A clerk of works is appointed by the architect or client to act as their on-site representative. On large contracts they will be resident on-site while on smaller contracts they will visit periodically. The COW is an 'inspector of works' and as such will ensure that the contractor carries out the work in accordance with the drawings and other contract documents. This includes inspecting both the standard of workmanship and the quality of materials. The COW will make regular reports back to the architect, keep a diary in case of disputes, make a daily record of the weather, of personnel employed on-site and any stoppages. The COW will also agree general matters directly with the building contractor. However the architect must confirm them to be valid.

Local authority

The local authority normally has the responsibility of ensuring that proposed building works conform to the requirements of relevant planning and building legislation. For this purpose, they employ planning officers and building control officers to approve and inspect building work. In some areas, building control officers are known as building inspectors or district surveyors (DS). Alternatively the client may appoint an approved inspector acting with the local authority, to approve the work in accordance with the Building Regulations and supervise during construction.

Health and safety inspector

The health and safety inspector, also known as the factory inspector, has the duty to ensure that the government legislation concerning health and safety is fully implemented by the building contractor.

Principal building contractor

The building contractor enters into a contract with the client to carry out, in accordance with the contract documents, certain building works. Each contractor will develop their own method and procedures for tendering and carrying out building work, which in turn, together with the size of the contract, will determine the personnel required. In addition they have the specific responsibility under the Construction (Design and Management) Regulations for health and safety onsite.

Subcontractors

The building contractor may call upon a specialist firm to carry out a specific part of the building work; for this they will enter into a subcontract, hence the term subcontractor. Contractor-appointed subcontractors may also be known as **domestic subcontractors**. The client or architect often names or nominates a specific subcontractor in the contract documents for specialist construction or installation work. These must be used for the work and are known as **nominated subcontractors**.

Suppliers

Building materials, equipment and plant are supplied by a wide range of merchants, manufacturers and hirers. The building contractor will negotiate with these to supply their goods in the required quantity and quality, at the agreed price, and finally in accordance with the building contractor's delivery requirements. The client or architect often nominates specific suppliers who must be used and are therefore termed **nominated suppliers**.

There is a recognized pattern by which the building team operates and communicates. This is best illustrated in the form of a line diagram, as shown below.

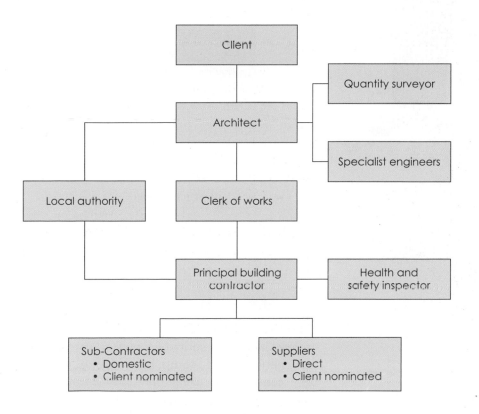

Figure 1.19

The building team

The building team may also be divided into a number of smaller teams, each with their specific interests and overlapping roles in the total building process.

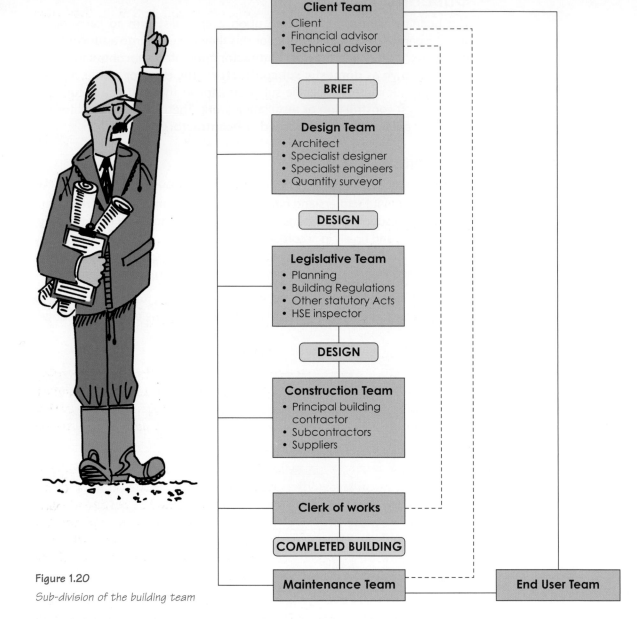

Figure 1.20

Sub-division of the building team

Building contractor

Building contractors may operate as **traditional companies**, directly employing the majority of their workforce to undertake the construction, or **management companies**, taking on a coordinating role during the construction using only subcontractors and suppliers, with many contractors operating some way between the two.

Traditionally building contractors only subcontracted out certain work such as structural steelwork, formwork, mechanical services and electrical installations, plastering, tiling and often painting. Today the move is towards a greater use of subcontractors for both the main and specialist operations, leaving the main contractor to perform a management role.

Subcontractors may be labour-only where they contract to fit the building contractor's material, or they may contract to supply and fix their own material.

Legal status

Whatever the size, method of operation or range of work undertaken by a building contractor it has to have a **legal status**, in one of the following categories:

◆ *Sole trader* – one person owns the business and organizes all the work.

◆ *Partnership* – two or more people own the business and share the responsibility of organizing the work.

◆ *Limited company (private)* – the business is registered with Companies House. Its owners are shareholders in the company. The main benefit to the owners is the fact that they are not personally liable for any business debts over the amount of their investment in the company.

◆ *Public Limited Company (plc)* – these are the largest companies; they are allowed to trade their shares on the stock market.

The building contractor's team

Contractors will develop their own method and procedures for estimating and carrying out building work, which, together with the size and nature of the work, will determine the number of personnel required. A typical medium to large building contractor's organizational structure is illustrated overleaf.

◆ *Estimator* – arrives at an overall cost for carrying out a building contract. In order to arrive at the overall cost they will break down each item contained in the bill of quantities into its constituent parts (labour, materials and plant) and apply a rate to each, representing the amount it will cost the contractor to complete the item. Added to the total cost of all items will be a percentage for overheads (head/site office costs, site management/administration salaries) and profit.

◆ *Buyer* – responsible for the purchase of materials; they will obtain quotations, negotiate the best possible terms, order the materials and ensure that they arrive on site at the required time, in the required quantity and quality.

◆ *Building contractor's quantity surveyor* – the building contractor's building economist; they will measure and evaluate the building work carried out each month, including the work of any subcontractors. An interim valuation is prepared on the basis of these measurements and passed on to the client for payment. They are also responsible for preparing interim costings to see whether or not the contract is within budget; finally they will prepare and agree the final accounts on completion of the contract.

◆ *Planning engineer* – responsible for the pre-contract planning of the building project; they plan the work in such a way as to ensure the most efficient/economical use of labour, materials, plant and equipment. Within

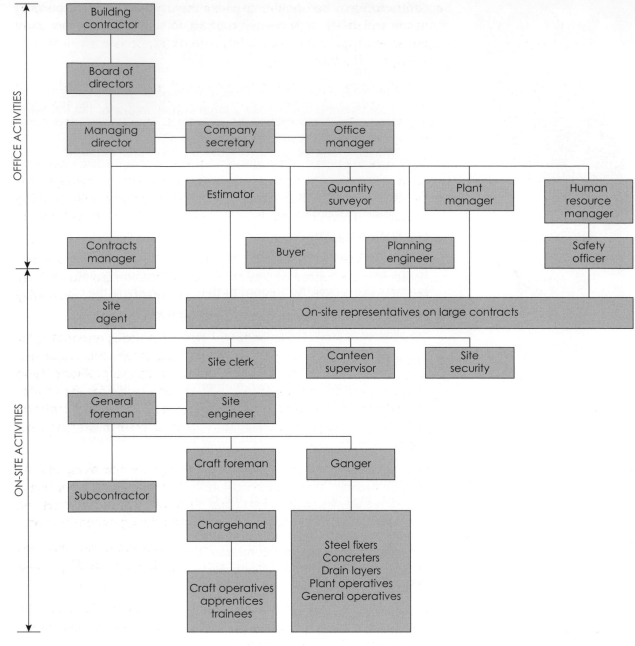

Figure 1.21 *Organizational structure: traditional medium to large building contractor*

their specialist field of work planning engineers are often supported by a work study engineer (to examine various building operations to increase productivity) and a bonus surveyor (to operate an incentive scheme, which is also aimed at increasing productivity by awarding operatives additional money for work completed over a basic target).

◆ ***Plant manager*** – responsible for all items of mechanical plant (machines and power tools) used by the building contractor. At the request of the contracts manager/site agent they will supply, from stock, purchase or hire, the most suitable plant item to carry out a specific task. They are also responsible for the maintenance of plant items and the training of operatives who use them.

Figure 1.22

◆ **Safety officer** – responsible to senior management for all aspects of health and safety. They advise on all health and safety matters, carry out safety inspections, keep safety records, investigate accidents and arrange staff safety training.

Each of the head office personnel mentioned above is the leader of a specialist service section and, depending on the size of the firm, will employ one or more technicians for assistance. On very large contracts there may also be a representative resident on-site.

◆ **Contracts manager** – the supervisor/coordinator of the site's management team, on a number of contracts. The contracts manager has an overall responsibility for planning, management and building operations. They will liaise between the head office staff and the site agents on the contracts for which they are responsible.

◆ **Site agent/site manager/project manager** – the building contractor's resident on-site representative and leader of the site workforce. They are directly responsible to the contracts manager for day-to-day planning, management and building operations.

◆ **General foreman** – works under the site agent and is responsible for coordinating the work of the craft foreman, ganger and subcontractors. They will also advise the site agent on constructional problems, liaise with the clerk of works and may also be responsible for the day-to-day employing and dismissing of operatives. On smaller contracts which may not require a site agent the general foreman will have total responsibility for the site.

◆ **Site engineer** – sometimes called the surveyor, works alongside the general foreman. They are responsible for ensuring that the building is the correct size and in the right place. They will set out and check the line, level and vertical (plumb) of the building during its construction.

◆ **Craft foreman** – works under the general foreman to organize and supervise the work of a specific craft, e.g. foreman bricklayer and foreman carpenter.

◆ **Ganger** – like the craft foreman, the ganger also works under the general foreman but is responsible for the organization and supervision of the general building operatives.

◆ **Chargehand/working foreman** – on large contracts employing a large number of craft operatives in each craft (normally bricklayers and carpenters), chargehands are often appointed to assist the craft foreman and supervise a subsection of the work, and also carry out the skilled physical work of their craft. For example, a foreman carpenter may have chargehands to supervise the carcassing team (floor joists and roofs); the first fixing team (flooring, frames and studwork); the second fixing team (doors, skirting, architraves and joinery fitments).

◆ **Operative** – the person who carries out the actual physical building work. Operatives can be divided into two main groups: **craft operatives** are the skilled craftsmen who perform specialist tasks with a range of materials, e.g. bricklayer, carpenter, electrician, painter, plasterer and plumber; building operatives are further subdivided into **general building operatives**, who mix concrete, lay drains, off-load material and assist craft operatives; and specialist building operatives, e.g. ceiling fixer, glazier, plant mechanic and scaffolder.

◆ *Site clerk* – responsible for all site administrative duties and the control of materials on-site. They will record the arrival and departure of all site personnel, prepare wage sheets for head office, record the delivery and transfer of plant items, record and check delivery of materials and note their ultimate distribution (assisted by a storekeeper).

8 Name and describe the role of **FOUR** members of the building team.

9 List the persons who would be employed in the design of a building.

10 Name the person who is responsible for checking the standard of work on behalf of the client on-site.

11 Name the person who is responsible for protection against injury on-site.

12 Name **THREE** craft operatives and give an example of the work **EACH** carries out.

13 Explain the term 'general operative'.

14 Explain the term 'subcontractor'.

15 A bill of quantities for building works is normally prepared for a client by the:
   a) building control officer
   b) clerk of works
   c) estimator
   d) quantity surveyor.

16 Name the job title of the person who controls a number of building contracts.

17 List **TEN** occupations of people employed by a building contractor.

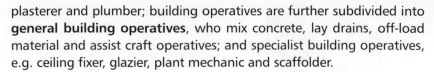

Principles of construction

A building encloses space and in doing so creates an **internal environment**. The actual structure of a building is termed the **external envelope**. This protects the internal environment from the outside elements, known as the **external environment**.

Types of building

The protective role of the building envelope is to provide the desired internal conditions for the building's occupants with regard to security, safety, privacy, warmth, light and ventilation (see Figure 1.23).

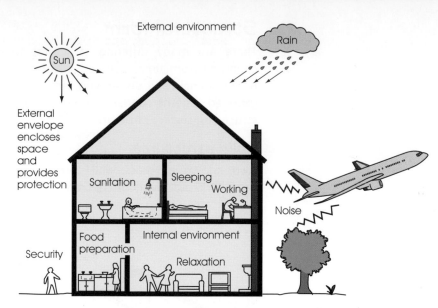

Figure 1.23

Internal/external environment

A **structure** or **construction** can be defined as an organized combination of connected elements (components), which are constructed and interconnected to perform some required function, e.g. a bridge. The term 'building' takes this idea a step further and is used to define structures that include an external envelope.

Buildings may be divided into three types according to their height, as illustrated in Figure 1.24.

◆ *Low-rise buildings* – from one to three storeys.

◆ *Medium-rise buildings* – from four to seven storeys.

◆ *High-rise buildings* – those above seven storeys.

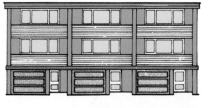

Figure 1.24

Height of buildings

High rise (over seven storeys)   Medium rise (four to seven storeys)   Low rise (one to three storeys)

These categories are further subdivided into a wide variety of basic shapes, styles and groupings, for example:

◆ *Detached* – a building that is unconnected with adjacent ones.

◆ *Semi-detached* – a building which is joined to one adjacent building but is detached from any other. It will share one dividing or party wall.

◆ *Terraced* – a row of three or more adjoining buildings, the inner ones of which will share two party walls.

Detached

Semi-detached

Terraced

Figure 1.25

The construction industry **Chapter 1**

Structural form

There are many differing structural forms in present-day use, each changing from time to time, in order to make the best possible use of new materials and developing techniques. These differing forms may be grouped together under three main categories: solid structures; framed structures; surface structures.

Solid structures

Also known as mass wall construction, these are constructed from brickwork, blockwork or concrete (see Figure 1.26).

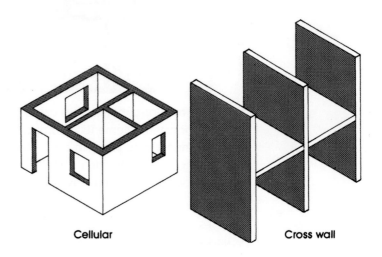

Cellular   **Cross wall**

Figure 1.26

Solid or mass wall structures

They form a stable box-like structure, but are normally limited to low-rise, short-span buildings.

Framed structures

Also termed skeleton construction, these consist of an interconnected framework of members having a supporting function (see Figure 1.18). Either external cladding or infill walls are used to provide the protecting external envelope. Frames made from steel, concrete or timber are often pre-made in a factory as separate units, which are simply and speedily erected on site. Framed construction is suitable for a wide range of buildings and civil engineering structures from low to high rise.

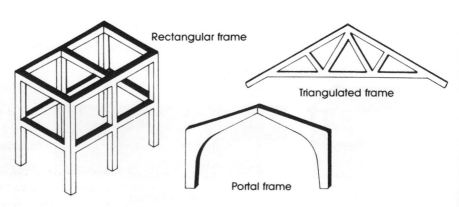

Rectangular frame

Triangulated frame

Portal frame

Figure 1.27

Frame structures

Surface structures

Consist of a thin material that has been curved or folded to obtain strength, or alternatively a very thin material that has been stretched over supporting members or medium. Surface structures are often used for large, clear span buildings with a minimum of internal supporting structure.

Structural parts

All structures consist of two main parts: that below ground and that above ground (see Figure 1.28).

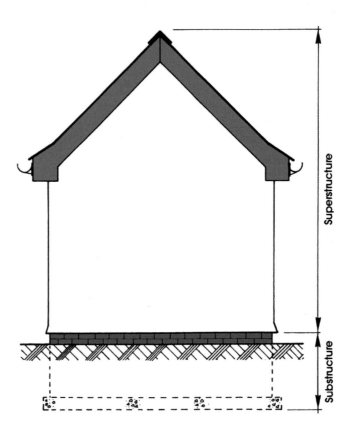

Figure 1.28
Structural parts

Substructure

This comprises all of the structure below ground and that up to and including the ground floor slab and damp-proof course. Its purpose is to receive the loads from the main building superstructure and its contents and transfer them safely down to a suitable load-bearing layer of ground.

Superstructure

This comprises all of the structure above the substructure both internally and externally. Its purpose is to enclose and divide space, and transfer loads safely on to the substructure. Although classified as separate parts, the substructure and the superstructure should be designed to operate as one structural unit.

Structural members and loading

The main parts of a structure, which themselves carry a load, are said to be in a state of stress (a body subjected to a force). There are three types of stress, as shown in Figure 1.29.

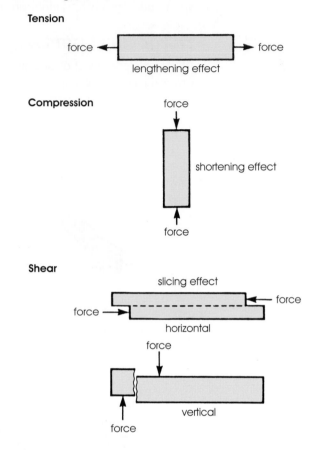

Figure 1.29

Types of stress

1. **Tension** – this tends to pull or stretch a material; it has a lengthening effect.

2. **Compression** – this causes squeezing, pushing and crushing; it has a shortening effect.

3. **Shear** – this occurs when one part of a member tends to slip or slide over another part; it has a slicing effect.

The two main **types of load** are defined as:

1. **Dead loads** – the self-weight of the building materials used in the construction, service installations and any permanent built-in fitments, etc.

2. **Imposed loads** – the weight of any movable load, such as the occupants of a building, their furniture and other belongings (goods and chattels), and any visitors to the property and their belongings. Also included are any environmental forces exerted on the structure from the external environment (wind, rain and snow).

The three main types of **load-bearing structural members** are illustrated in Figure 1.30.

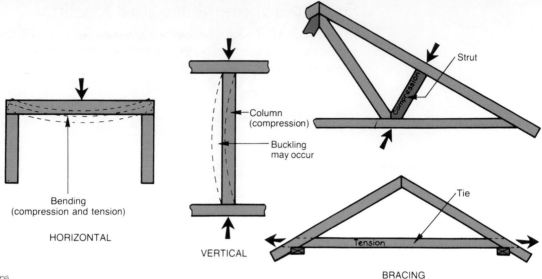

Figure 1.30

Types of load-bearing members

1. *Horizontal members* – their purpose is to carry and transfer a load back to its point of support. Horizontal members include beams, joists, lintels and floor or roof slabs, etc. When a load is applied to a horizontal member, bending will occur, resulting in a combination of tensile, compressive and shear stresses. This bending causes compression in the top of the member, tension in the bottom and shear near its supports and along its centre line. Bending causes the member to sag or deflect. (For safe design purposes deflection is normally limited to a maximum of 3 mm in every 1-metre span.) In addition, slender members, which are fairly deep in comparison with their width, are likely to buckle unless restrained (e.g. strutting to floor joists).

2. *Vertical members* – their purpose is to transfer the loading of the horizontal members down onto the substructure. Vertical members include walls, columns, stanchions and piers. Vertical members are in compression when loaded. Buckling tends to occur in vertical members if they are excessively loaded or are too slender.

3. *Bracing members* – they are used mainly to triangulate rectangular frameworks in order to stiffen them. These can be divided into two types: the strut, a bracing member that is mainly in compression; and the tie, a bracing member that is mainly in tension.

At certain times bracing members may, depending on their loading conditions, act as either struts or ties. In these circumstances they may be termed as **braces**.

Examples of loading

An example of dead and imposed loads and how they are transferred down through the structural members to the soil is illustrated in Figure 1.31.

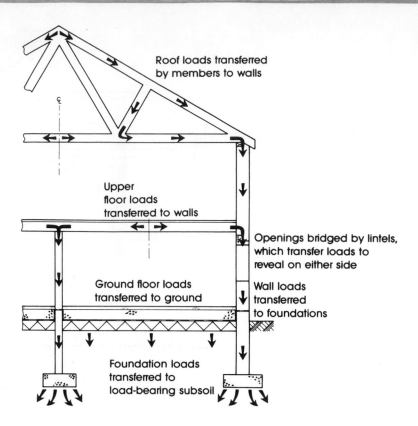

Roof loads transferred
by members to walls

Upper
floor loads
transferred to walls

Openings bridged by lintels,
which transfer loads to
reveal on either side

Ground floor loads
transferred to ground

Wall loads
transferred
to foundations

Foundation loads
transferred to
load-bearing subsoil

Figure 1.31

Transfer of loads

Building elements

An element can be defined as a constructional part of either the substructure or superstructure, having its own functional requirements. These include the foundations, walls, floors, roof, stairs and the structural framework or skin. Elements may be further classified into three main groups: primary, secondary and finishing elements.

Primary elements

These are named because of the importance of their supporting, enclosing and protection functions. In addition, they have mainly internal roles of dividing space and providing floor-to-floor access. Typical examples of primary elements are shown in Figure 1.32.

Primary element – foundations

This is the part of the structure (normally *in situ* concrete) that transfers the dead weight and imposed loads of the structure safely onto the ground. The width of a foundation is determined by the total load of the structure exerted per square metre on the foundation and the safe load-bearing capacity of the ground. Wide foundations are used for either heavy loads or weak ground and narrow foundations for light loading or high load-bearing ground. The load exerted on foundations is spread to the ground at an angle of 45 degrees. Shear failure leading to building subsidence (sinking) will occur if the thickness of the concrete is less than the projection from the wall/column face to the edge of the foundation. Alternatively, steel reinforcement may be included to enable the load to spread across the full width of the foundation (see Figure 1.33).

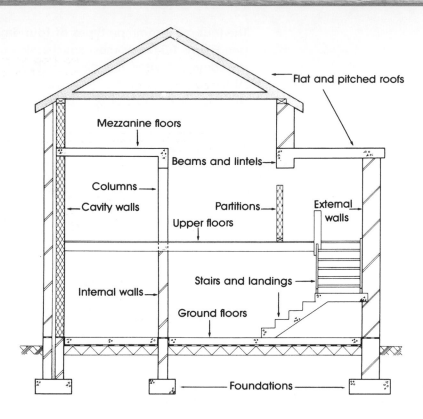

Figure 1.32

Primary elements

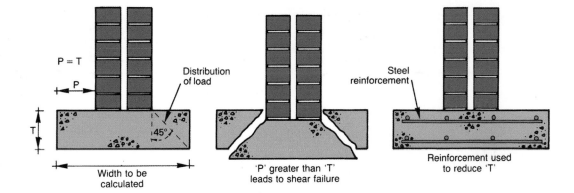

Figure 1.33

Foundation properties

Foundations are taken below ground level to protect the structure from damage resulting from ground movement. The actual depth below ground level is dependent on a number of factors: load-bearing capacity of the ground; need to protect against ground movement and tree roots, etc. In most circumstances, a depth of 1 metre to the bottom of the foundation is considered to be the minimum.

Ground movement is caused mainly by the shrinkage and expansion of the ground near the surface owing to the wet and dry conditions. Compact granular ground suffers little movement, whereas a clay (cohesive) ground is at high risk. Frost also causes ground movement when the water in the ground expands on freezing. This is known as frost heave and is limited to about 600 mm in depth. The main problem with tree roots is shrinkage of the ground owing to the considerable amounts of water they extract from it. Tree roots can extend out in all directions further than its height.

The four most common **types of foundations** are strip, pad, raft and pile (see Figure 1.34). For most small-scale building works **strip foundations** are commonly used for solid structures and **pad foundations** for frame structures, except where the subsoil is of a poor, unstable quality. In these circumstances, a **raft** or **pile foundation** would be more suitable.

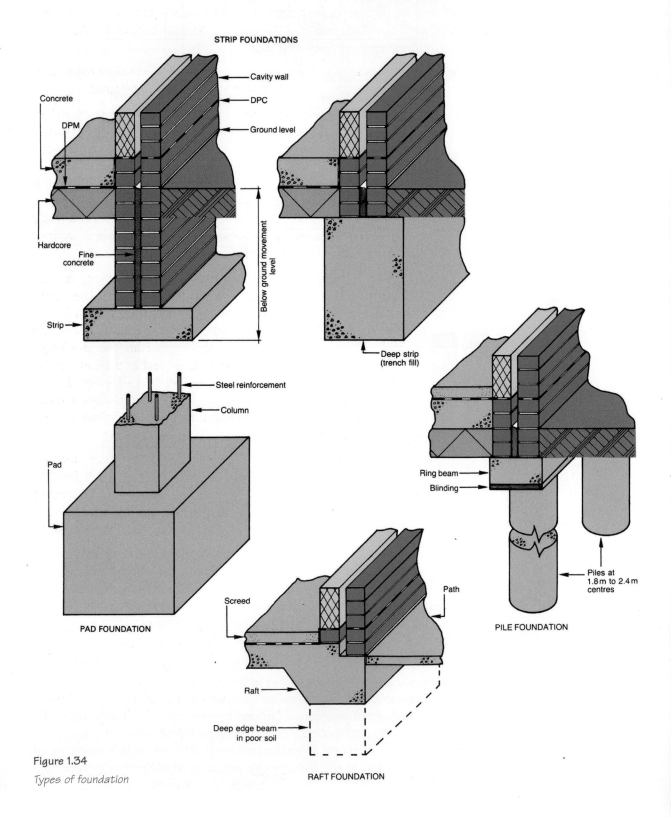

Figure 1.34

Types of foundation

Primary element – walls

The walls of a building may be classed as either load bearing or non-load bearing. In addition, external walls have an enclosing role and internal walls have a dividing one. Thus load-bearing walls carry out a dual role of supporting and enclosing or dividing. Internal walls, both load and non-load bearing are normally termed **partitions**. Openings in load-bearing walls (windows and doors) are bridged by either arches or lintels, which support the weight of the wall above.

Walls may be divided into three main groups according to their method of construction (see Figure 1.35). These are solid, cavity and framed:

◆ *Solid walls* – these are made from bricks, blocks or concrete. When used externally, very thick walls are required (450 mm or over) in order to provide sufficient thermal insulation. This thickness also prevents rain being absorbed through the wall to the inside causing internal dampness before heat and air circulation can evaporate it from the outside. Because of the costs involved this method is now rarely used. An alternative method is to use thinner external solid walls (normally insulating blocks) and to apply an impervious (waterproof) surface finish to the outside, e.g. cement rendering.

◆ *Cavity walls* – these consist of two relatively thin walls or 'leaves' (about 100 mm each) separated by a 50–75-mm cavity. The cavity prevents the transfer of moisture from the outside to the inside and also improves the wall's thermal insulation properties. Cavity walls are in common use for enclosing walls of low to medium-rise dwellings. The standard form of the wall is a brick outer leaf and an insulating block inner leaf or, as an alternative, a timber-framed inner leaf. To reduce heat transfer through the wall, it is fairly common practice to fill the cavity with a **thermal insulating material** (mineral wool, fibreglass, foam or polystyrene, etc.).

◆ *Framed walls* – these are normally of timber construction and are made up in panels. They may be either load or non-load bearing and for use externally or internally. They consist of vertical members or **studs** and horizontal members, the top and bottom of which are called the head and sole plates, while any intermediates are called

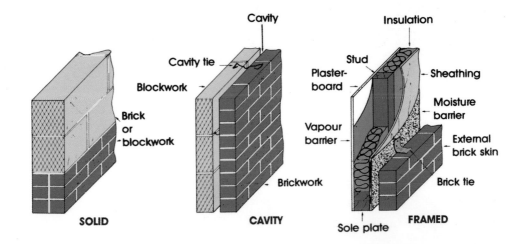

Figure 1.35

Types of wall

noggins. Their thermal and sound insulating properties are greatly improved by filling the spaces between the members with mineral wool or fibreglass, etc. Sheathing may be fixed to one or both sides of the panel to improve strength. This type of wall panel is used in the majority of present-day timber-frame house construction as the internal 'leaf' of the cavity wall.

The strength of brickwork of solid and cavity walls is dependent on its bonding (overlapping of vertical joints). This is necessary to spread any loading evenly throughout the wall (see Figure 1.36). The actual overlap or bonding pattern will vary depending on the type of wall and the decorative effect required; typical examples are illustrated in Figure 1.37.

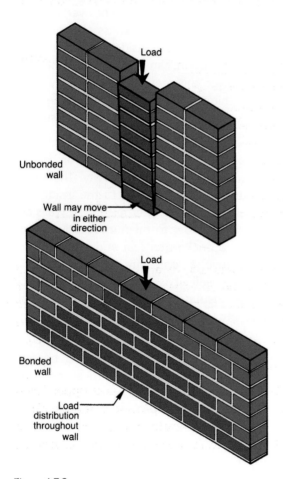

Figure 1.36

Reason for bonding

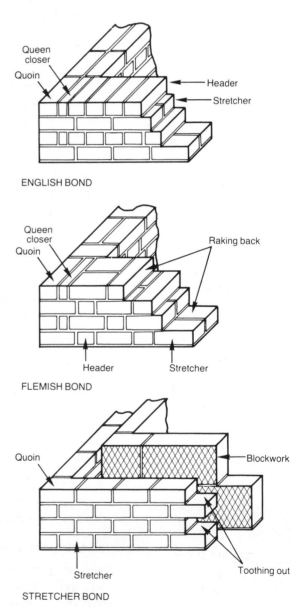

Figure 1.37

Types of bond

◆ Solid walls are normally built in either English or Flemish bond. **English bond** consists of alternate rows (courses) of bricks laid lengthways along the wall (stretchers) and bricks laid widthways across the wall (headers). **Flemish bond** consists of alternate stretchers and headers in the same course. In both, the quarter lap is formed by placing a queen closer (brick reduced in width) next to the quoin (corner brick).

◆ Cavity walls are built in **stretcher bond** where all bricks show their stretcher faces, although adjacent courses overlap by half a brick. To ensure sufficient strength, the inner and outer leaves are tied together across the cavity at intervals with cavity ties.

Openings in walls for doors and windows are spanned by steel or concrete lintels. These bridge the opening and transfer loads to the reveal on either side (see Figure 1.38).

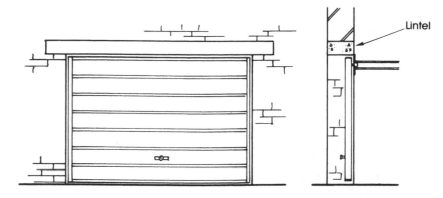

Figure 1.38 *Lintel*

Brickwork and blockwork walls are jointed by means of mortar (mixture of sand and cement and/or lime forming an adhesive). Horizontal joints are known as **bed joints** and vertical ones as **perpends** or perps. The face of these joints may be finished in a variety of profiles intended to improve the weathering resistance and appearance of the work (see Figure 1.39).

Primary element – floors

These are the horizontal internal surfaces at ground and upper levels (see Figure 1.40). Their main functions are to provide a level surface, standard of insulation and carry and transfer any loads imposed upon it. In addition, ground floors are also required to prevent moisture penetration and weed growth.

◆ Ground floors are either solid or hollow (suspended).
◆ Upper floors are suspended: timber construction is mainly used for house construction and concrete for other works.

Primary element – roofs

These are part of the external envelope that spans the building at high level and has weathering and insulation functions. They are classified according to their **roof pitch** (slope of the roof surface) and also their shape, the most common of which are illustrated in Figure 1.41.

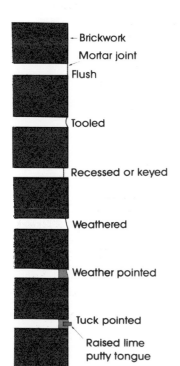

Figure 1.39

Joining brickwork

The construction industry · **Chapter 1**

Figure 1.40

Floors

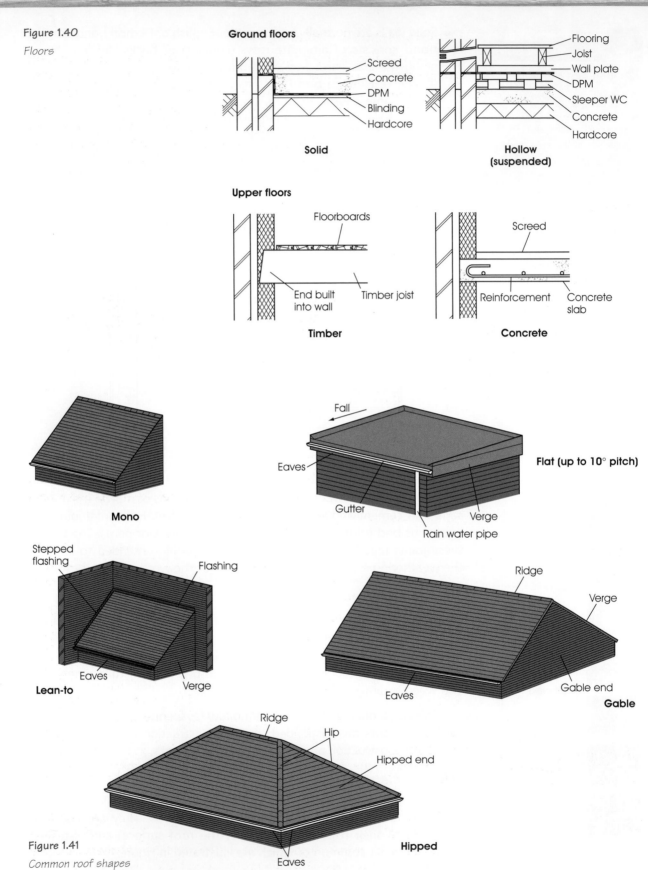

Ground floors

Solid:
- Screed
- Concrete
- DPM
- Blinding
- Hardcore

Hollow (suspended):
- Flooring
- Joist
- Wall plate
- DPM
- Sleeper WC
- Concrete
- Hardcore

Upper floors

Timber:
- Floorboards
- End built into wall
- Timber joist

Concrete:
- Screed
- Reinforcement
- Concrete slab

Mono

Flat (up to 10° pitch)
- Fall
- Eaves
- Gutter
- Verge
- Rain water pipe

Lean-to
- Stepped flashing
- Flashing
- Eaves
- Verge

Gable
- Ridge
- Verge
- Eaves
- Gable end

Hipped
- Ridge
- Hip
- Hipped end
- Eaves

Figure 1.41

Common roof shapes

Primary element – stairs

Stairs provide floor-to-floor access. They can be defined as a series of steps (combination of tread and riser), each continuous set of steps being called a flight. Landings may be introduced between floor levels, to break up a long flight giving rest points, or to change the direction of the stair. Stairs can be classified according to their plan shape as shown in Figure 1.42 or by the material from which they are made. Timber stairs are common in dwelling houses, while concrete is most common for other works.

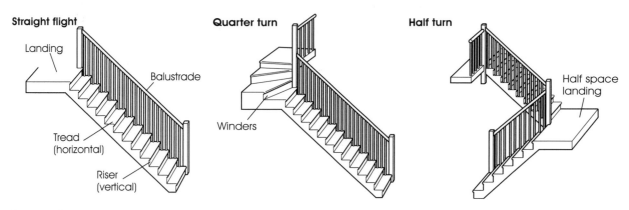

Figure 1.42

Classification of stairs by plan shape

Secondary elements

Secondary elements are the non-essential elements of a structure, having mainly a completion role around openings in primary elements and within the building in general; see Figure 1.43.

Figure 1.43

Secondary elements

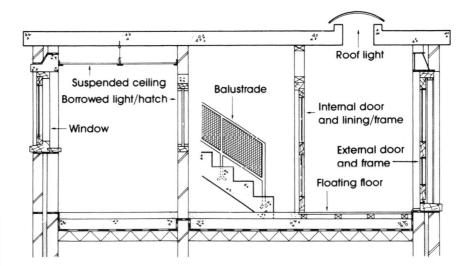

◆ **Doors** are movable barriers used to cover an opening in a structure. Their main function is to allow access in a building and passage between its interior spaces. Other functional requirements include weather protection, fire resistance, sound and thermal insulation, security, privacy, ease of operation and durability. They may be classified by their method of construction and method of operation.

The construction industry

Chapter 1

The surround to the wall opening on which doors are hung may be either a frame or lining.

◆ **Windows** are glazed openings in a wall used to allow daylight and air in and give occupants a view outside. Windows are normally classified by their method of opening and the material from which they are made. Windows projecting beyond the face of a building at ground level are bay windows; those projecting from an upper storey are oriel windows; those with a continuous curve are bow windows; those that contain a pair of casements for giving access to a garden or balcony are French windows.

Finishing elements

A finish is the final surface of an element, which can be a self-finish as with face brickwork and concrete or an applied finish such as plaster, wallpaper and paint. Typical examples of finishing elements are shown in Figure 1.44. Included in this category are internal trims (skirting, architraves and coving or cornices), which mask the joint between adjacent elements, external flashings that weatherproof the joint and cladding, cement rendering and tile hanging, which are all used to either weatherproof or give a decorative finish to external walls.

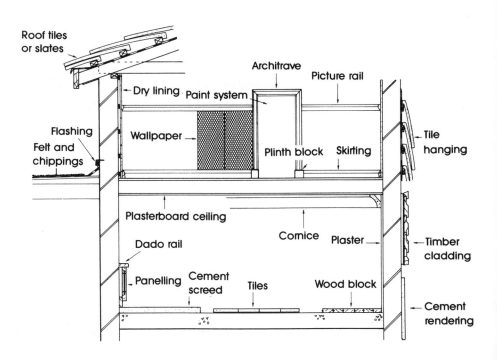

Figure 1.44

Finishing elements

Building components and services

The primary elements, secondary elements and services of a building are made up from a number of different parts or materials; these are known as components. Examples of three main types of components are shown in Figure 1.45:

◆ *Section components* – section is a material that has been processed to a definite cross-sectional size but of an unspecified or varying length, e.g. a length of timber.

◆ *Unit components* – unit is a material that has been processed to a definite cross-sectional size and length, e.g. a brick.
◆ *Compound components* – these are combinations of sections or units put together to form a complex article, e.g. a window frame.

The various components are combined to form the elements of a building.

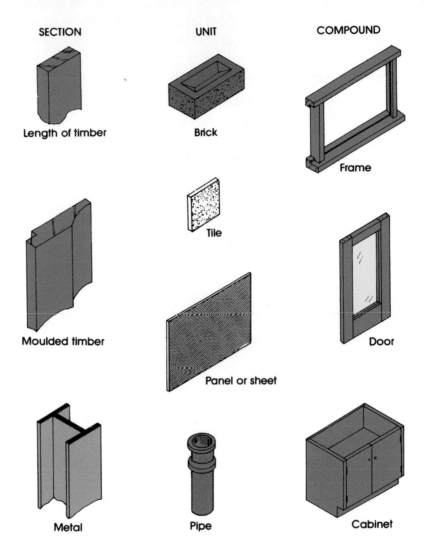

SECTION   UNIT   COMPOUND

Length of timber

Brick

Frame

Moulded timber

Tile

Door

Panel or sheet

Metal   Pipe   Cabinet

Figure 1.45
Building components

Building services

Certain basic services are considered as essential requirements for all buildings:

◆ **Water** for drinking, washing, heating, cooking, waste/soil disposal and industrial processes (Figure 1.46).
◆ **Drainage** for the disposal of wastewater and sewage (Figure 1.47).
◆ **Electricity** for lighting, heating, cooking, cleaning, air conditioning, entertainment, telecommunications and industrial processes (Figure 1.48).
◆ **Gas** for heating, cooking and industrial processes (Figure 1.49).

(a)

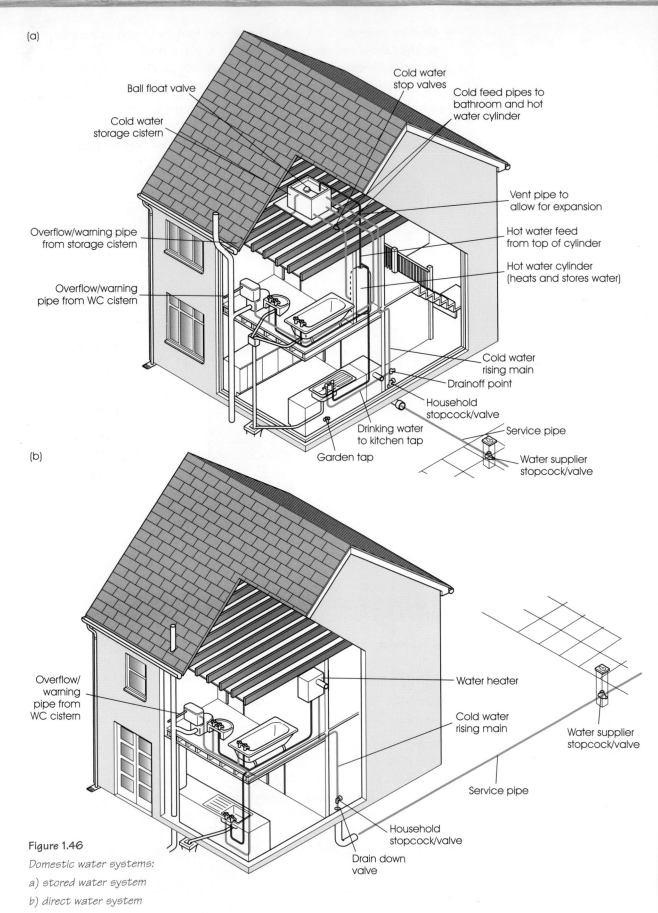

Ball float valve

Cold water stop valves

Cold feed pipes to bathroom and hot water cylinder

Cold water storage cistern

Vent pipe to allow for expansion

Overflow/warning pipe from storage cistern

Hot water feed from top of cylinder

Overflow/warning pipe from WC cistern

Hot water cylinder (heats and stores water)

Cold water rising main

Drainoff point

Household stopcock/valve

Drinking water to kitchen tap

Service pipe

Garden tap

Water supplier stopcock/valve

(b)

Overflow/ warning pipe from WC cistern

Water heater

Cold water rising main

Water supplier stopcock/valve

Service pipe

Household stopcock/valve

Drain down valve

Figure 1.46

Domestic water systems:

a) stored water system

b) direct water system

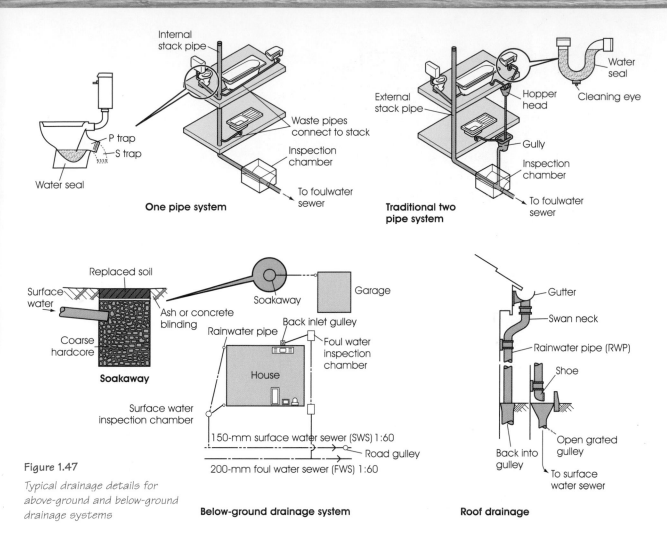

Figure 1.47

Typical drainage details for above-ground and below-ground drainage systems

did you know?

Competent persons
are those who have the experience, knowledge and appropriate qualifications to carry out their work role. Those working on service installations must, by law, be registered with the appropriate regulatory body.

These basic services consist of systems of pipes or wires, which are either fixed within, or on, the surface of the elements. They are connected to the distribution system, usually via a meter, which records the amount used.

Each supply company has its own set of regulations concerning the supply, use of, and any alterations to, its services. These regulations must be complied with when provision is made within a building for connection to the particular service.

The installation and maintenance of services should be undertaken only by **competent persons**.

The construction industry

Chapter 1

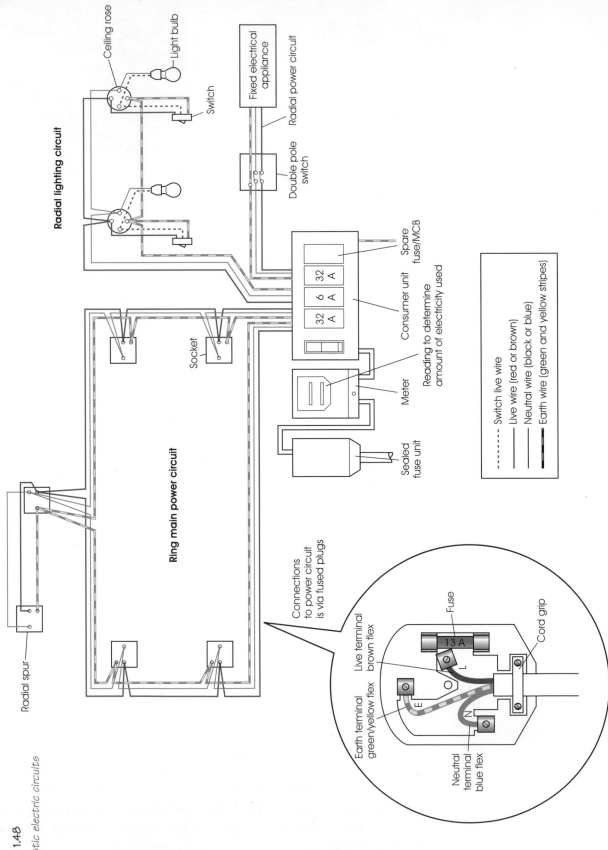

Figure 1.48

Domestic electric circuits

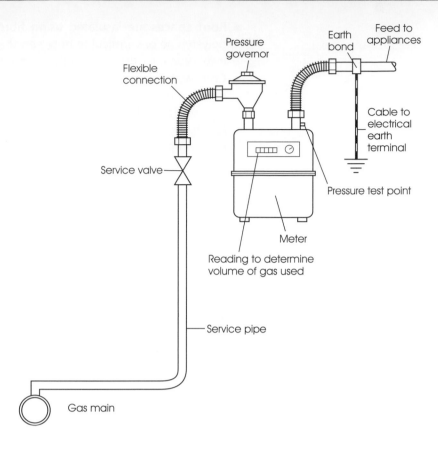

Figure 1.49

Domestic gas supply

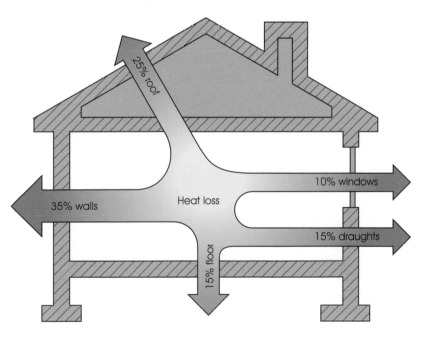

Figure 1.50

Typical heat loss through building elements

Thermal insulation

Thermal insulation is essential if heating costs are not to be wasted. Figure 1.50 illustrates the percentage heat losses through the various building elements of a typical house.

◆ **Roof spaces** are insulated using fibreglass, mineral wool, expanded polystyrene or vermiculite between the ceiling joists. Laying additional insulation over the ceiling joists can top up existing insulation.

◆ **Cold-water storage tanks** should be lagged to prevent them freezing in the winter. As an extra precaution, the tank will be kept warmer if the ceiling insulation is not continued under them.

◆ **Cavity walls**, where not already filled with insulation during the building process, can have insulating material injected into them. This may be plastic foam, polystyrene granules or mineral wool fibre.

◆ **Solid walls** can be improved by adding an internal insulating layer. This may take the form of a cavity by lining the walls with foil-backed plasterboard on battens. Extra benefit is achieved by filling the space between battens with fibreglass or mineral wool.

◆ **Sealed-unit double-glazing** is most effective against heat loss through windows and glazed doors.

◆ **Secondary double-glazing** is normally not as good as sealed units since its primarily use is for sound insulation.

◆ **Hollow ground floors** may be improved by laying fibreglass or mineral wool between the joists and suspended on wire mesh. A floating floor finish with insulation between the battens may be used to improve **solid ground floors**.

◆ **Flexible sealing strips** are a simple and fairly cheap method of cutting down on heat loss through draughts around the joints of windows and doors and over postal flaps.

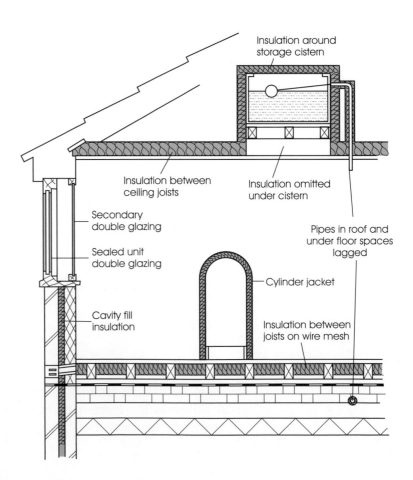

Insulation around storage cistern

Insulation between ceiling joists

Insulation omitted under cistern

Secondary double glazing

Sealed unit double glazing

Pipes in roof and under floor spaces lagged

Cylinder jacket

Cavity fill insulation

Insulation between joists on wire mesh

Figure 1.51

Maximum for domestic buildings

Site works

Excavation

The process of removing earth, to form a hole in the ground, can be dug manually using a shovel or mechanically using a digger excavator.

Oversite excavation is the removal of topsoil and vegetable matter from a site prior to the commencement of building work. Excavation depth is typically between 150 and 300 mm.

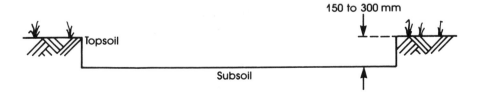

Figure 1.52

Oversite excavation

Reduced level excavation is a process carried out after oversite excavation on undulating ground. It consists of cutting and filling operations to produce a level surface, called the formation level.

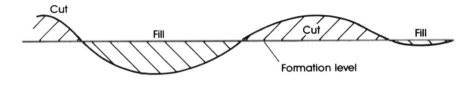

Figure 1.53

Reduced level excavation

Trench excavations are long, narrow holes in the ground, to accommodate strip foundations or underground services. Deep trenches may be battered or timbered to prevent the sides from caving in.

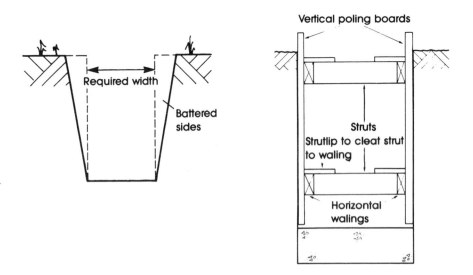

Figure 1.54

Timbering to sides of trench

Pit excavations are deep rectangular holes in the ground, normally for column base pad foundations. Larger holes may be required for basements etc. Sides may be battered or timbered depending on depth.

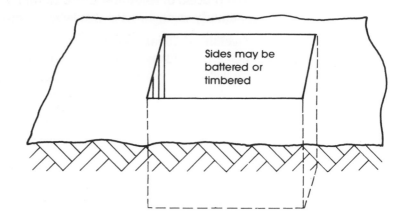

Figure 1.55

◆ **Road construction** – in this context the scope is limited to small estate roads, access roads and driveways. Once excavated and scraped down to the formation level construction of the surface can commence on the sub-grade.

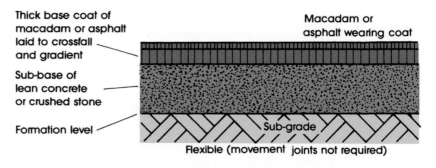

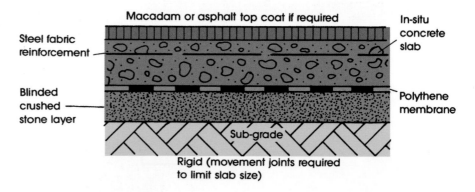

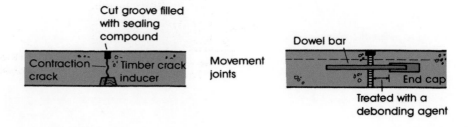

Figure 1.56

◆ *Steel fixing* – concrete is very strong when being compressed (squashed) but comparatively weak in tension (stretched or bent). Few structures are subjected to loadings that are totally compressive. Thus steel reinforcement is normally introduced to increase strength and prevent structural failure, thereby producing a composite material called reinforced concrete.

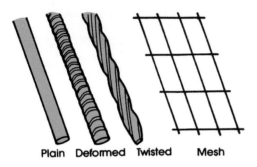

Figure 1.57

Steel reinforcement

Shape	Code	Type
	00	Straight
	13	Hook end
	11	Straight turned up one end
	21	Turned up both ends
	26	Cranked for shear
	51	Stirrups or links

Figure 1.58

Typical standard shape codes

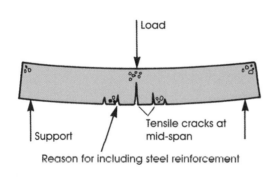
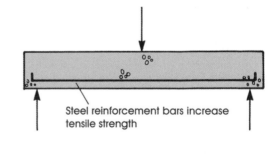

Figure 1.59

Concrete cover to steel reinforcement prevents corrosion of steel

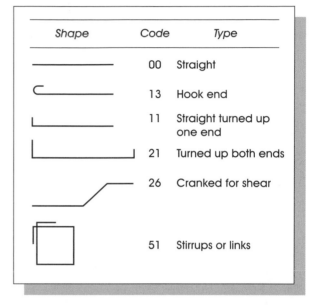
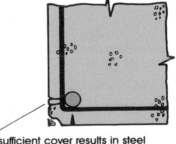

◆ *Formwork* – a structure that is usually temporary but can be partly or wholly permanent, designed to contain fresh, fluid concrete, form it into the required shape and dimension and support it until it cures (hardens) sufficiently to become self-supporting. The surface in contact with the concrete is known as the form face while the supporting structure can be referred to as falsework.

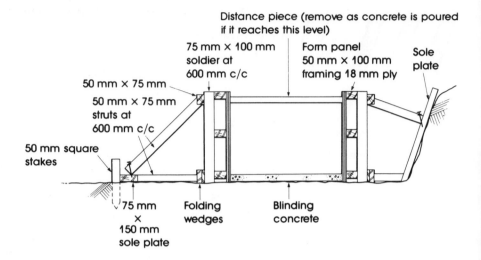

Figure 1.60

Column pad base formwork

18. Sketch **TWO** types of building structure.

19. Name **TWO** types of foundation.

20. State the purpose of steel reinforcement in concrete.

21. Sketch and label a typical cavity wall.

22. Sketch and label a typical timber upper floor construction.

23. State the purpose of formwork.

24. Name **TWO** methods of excavation.

25. List **FOUR** finishing elements.

26. State **TWO** types of road surface.

The illustration below shows a part section through a single-storey building, to which the following questions relate:

Define the following terms and indicate an example of each on the building section. You may photocopy the illustration or download it from the web at www.nelsonthornes.com/carpentry.

1. Substructure
2. Superstructure
3. Primary element
4. Secondary element
5. Finishing element
6 Component.

Name the type of foundation and wall construction.

Identify the lettered elements/components shown in the secton and name the material indicated.

Describe three measures that could be undertaken that would improve the thermal insulation of the building.

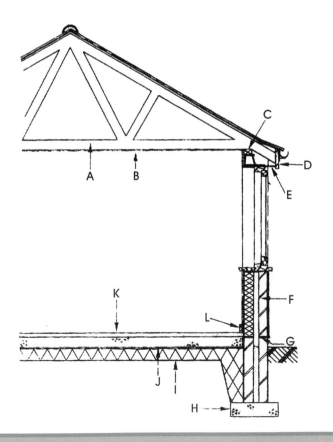

Figure 1.61

Building section

Health and Safety

This chapter is intended to provide the new entrant with an overview of health and safety issues in the construction industry. Its contents are assessed in the **NVQ Unit No. VR 01 Conform to General Workplace Safety**. It is concerned with safety legislation and the general safety issues that must be complied with in the workplace on a day-to-day basis.

In this chapter you will cover the following range of topics:

◆ Health and safety overview
◆ Emergency procedures
◆ Accidents
◆ Health and safety controls
◆ The Health and Safety at Work Act (HASAWA)
◆ Management of health and safety legislation
◆ Other safety regulations
◆ Regulation of hazardous substances
◆ General site safety
◆ Site security.

Health and safety overview

did you know?

Hazards are something with the potential to cause harm.

Harm can vary in its severity; some hazards can cause death, others illness or disability or maybe only cuts or bruises.

Risk is concerned with the severity of harm and the likelihood of it happening.

Health and safety forms an essential part of your daily working life. Construction site safety is paramount for the well-being of all concerned. Ensuring that a site is as safe as it can possibly be is a shared responsibility between employers and the workforce. Employers have a duty to create safe working conditions and provide the workforce with training, which explains safety rules, procedures and regulations. You as part of the workforce have a major contribution to make in site safety, by responding to safety instructions, complying with safety rules and developing the skills to identify potential safety hazards and reduce risks.

Emergency procedures

Emergencies are situations or events that require immediate action; examples of emergencies include fire, accidents, bomb threat, leakage of chemicals or other hazardous substance.

Although many of these examples are fairly rare it is essential that individuals are aware of the procedures and responsibilities for dealing with them.

Procedures

All personnel on-site should be aware of the following:

◆ The type of evacuation alarm, e.g. siren, tannoy or load hailer etc.
◆ On the sounding of an alarm, you must proceed to your designated assembly point
◆ Whom you have to report to (name of nominated person) on arrival at the assembly point
◆ What are the procedures to be followed after reporting to the nominated person
◆ The emergency services (fire, police or ambulance) will advise the nominated person when the emergency is over
◆ You must remain at the assembly point until advised otherwise by the nominated person
◆ Under no circumstances should you re-enter the site or building until the nominated person authorizes it.

Roles and responsibilities

Your role as part of the workforce in dealing with emergencies is restricted to the following:

◆ Raise alarm or report details of emergency to the nominated person
◆ Report potential hazards and any near misses to the nominated person
◆ Ensure that your own safety is not at risk
◆ Follow the emergency procedures as instructed
◆ Assist in the completion of accident records and reports
◆ If you are first on the scene you may also be required to:
  ▸ Make a 999 call to summon the emergency services if there is a fire, people are injured or if there is a threat to life
  ▸ Call for the first aider on-site or give emergency help yourself if you are a qualified first aider.

Making a 999 call

◆ Dial 999 on a landline or mobile phone
◆ Keep calm and listen to the operator's instructions
◆ State your name and the service you require when asked
◆ On connection to the required service, again remain calm and follow the operator's instructions

Health and safety

Chapter 2

◆ You will be asked to explain the nature of the emergency and its location; try to give exact details such as name of company, address and postcode

◆ On completion of the call arrange for someone to meet the the emergency servive to guide them to the incident.

Accidents

Definition

An accident is often described as a chance event or an unintentional act. This description is not acceptable. Accidents do not 'just happen', they do not 'come out of the blue', they are caused! A better definition of an accident is therefore:

> 'An event causing injury, damage or loss that might have been avoided by following correct methods and procedures.'

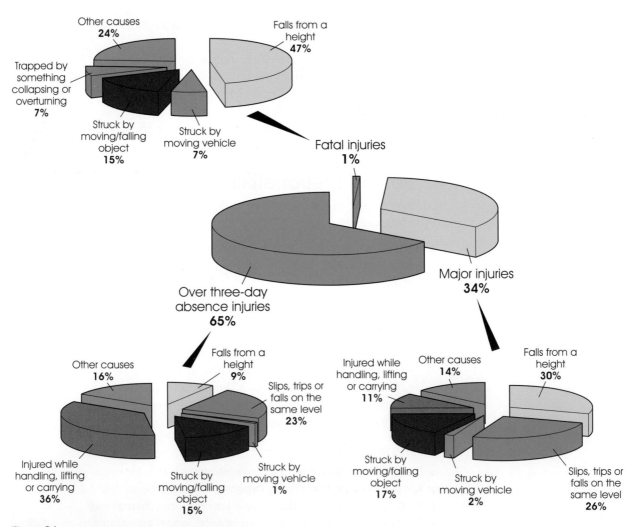

Figure 2.1

Distribution of reported accidents by type and cause

Accident statistics

Each year there are over 14,000 accidents reported to the Health and Safety Executive that occur during construction-related activities in the UK.

Reported accidents are those that result in death, major injury, more than three days' absence from work or are caused by a notifiable dangerous occurrence.

At the time of writing figures show that 1% of the reported accidents were fatal, on average 6 deaths a month; 34% resulted in major injuries, nearly 20 a week; and 65% resulted in absence from work for more than three days, an average of 40 each working day.

These percentage distributions along with a further breakdown by cause of accident are illustrated in the pie charts in Figure 2.1.

These figures are not intended to frighten you or put you off a career in the construction industry, but simply to make you aware of the hazards involved so that all concerned will make a conscious effort to improve them.

Figure 2.2

SAFETY CONSCIOUS PEOPLE = SAFE ACTIONS = SAFE WORKING CONDITIONS

measuring up

1. The greatest number of fatal accidents in the construction industry involve:
   a) machinery
   b) electric shock
   c) falls
   d) transport.

2. Define the term 'accident'.

3. Define what constitutes a reported accident and state to whom it is reported.

4. State **THREE** of your responsibilities as a member of the workforce, in the event of an emergency occurring on-site.

Health and safety

Chapter 2

Health and safety controls

In the mid-1970s the Health and Safety at Work etc. Act was introduced (HASAWA). HASAWA was seen as an enabling umbrella. It introduced the main statutory legislation, completely covering the health and safety of all persons at their place of work and protecting other people from risks occurring through work activities.

It has overseen the gradual replacement of previous piecemeal health and safety requirements by revised and up-to date measures prepared in consultation with industry and its workers. As a member of the construction industry you are required to know about your responsibilities with regard to various safety legislations.

The Health and Safety at Work Act (HASAWA)

The four main objectives of HASAWA are as follows:

1. To secure the health, safety and welfare of all persons at work
2. To protect the general public from risks to health and safety arising out of work activities
3. To control the use, handling, storage and transportation of explosives and highly flammable substances
4. To control the release of noxious or offensive substances into the atmosphere.

These objectives can be achieved only by involving everyone in health and safety matters, including:

- ◆ Employers and management
- ◆ Employees (and those undergoing training)
- ◆ Self-employed
- ◆ Designers, manufacturers and suppliers of equipment and materials.

Figure 2.3

Accidents!

Employers' and management duties

Employers have a general duty to ensure the health and safety of their employees, visitors and the general public. This means that the employer must:

1. Provide and maintain a safe working environment
2. Ensure safe access to and from the workplace

did you know?

An employer is not allowed to charge an employee for anything done, or equipment provided, to comply with any health and safety requirement.

3. Provide and maintain safe machinery, equipment and methods of work

4. Ensure the safe handling, transport and storage of all machinery, equipment and materials

5. Provide employees with the necessary information, instruction, training and supervision to ensure safe working

6. Prepare, issue to employees and update as required a written statement of the firm's safety policy

7. Involve trade union safety representatives (where appointed) with all matters concerning the development, promotion and maintenance of health and safety requirements.

Employees' duties

An employee is an individual who offers his or her skill and experience etc. to his or her employer in return for a monetary payment. It is the duty of all employees while at work to comply with the following:

1. Take care at all times and ensure that their actions do not put at 'risk' themselves, their workmates or any other person

2. Cooperate with their employers to enable them to fulfil the employers' health and safety duties

3. Use the equipment and safeguards provided by the employers

4. Never misuse or interfere with anything provided for health and safety.

Figure 2.4

SAFETY BEGINS
WITH YOU

Figure 2.5

Self-employed duties

The self-employed person can be thought of as both the employer and employee; therefore their duties under the Act are a combination of those of the employer and employee.

Designers', manufacturers' and suppliers' duties

Under the Act, designers, manufacturers and suppliers as well as importers and hirers of equipment, machinery and materials for use at work have a duty to:

1. Ensure that the equipment machinery or material is designed, manufactured and tested so that when it is used correctly no hazard to health and safety is created

Health and safety   Chapter 2

WARNING
CONTAINS
ASBESTOS

Breathing asbestos
dust is dangerous
to health

Follow safety
instructions

Figure 2.6

did you know?

Employers should ensure
this information is passed on
to their employees.

did you know?

The HSE section with the
main responsibility for the
building industry is the
Factory Inspectorate.

2. Provide information or operating instructions as to the correct use, without risk, of their equipment, machinery or material.

3. Carry out research so that any risk to health and safety is eliminated or minimized as far as possible.

Enforcement of safety legislation

Under HASAWA a system of control was established, aimed at reducing death, injury and ill health, by means of the Health and Safety Executive (HSE). The Executive is divided into a number of specialist inspectorates or sections, which operate from local offices situated throughout the country. From the local office, inspectors visit individual workplaces.

Health and Safety Executive inspectors have been given wide powers of entry, examination and investigation in order to assist them in the enforcement of HASAWA and other safety legislation. In addition to giving employers advice and information on health and safety matters, an inspector can do the following:

1. **Enter premises** in order to carry out investigations, including the taking of measurements, photographs, recordings and samples. The inspector may require the premises to be left undisturbed while investigations are taking place.

2. **Take statements**. An inspector can ask anyone questions relevant to the investigation and also require them to sign a declaration as to the truth of the answers.

3. **Check records**. All books, records and documents required by legislation must be made available for inspection and copying.

4. **Give information**. An inspector has a duty to give employees or their safety representative information about the safety of their workplace and details of any action they propose to take. This information must also be given to the employer.

5. **Demand**. The inspector can demand the seizure, dismantling, neutralizing or destruction of any machinery, equipment, material or substance that is likely to cause immediate serious personal injury.

6. **Issue an improvement notice**. This requires the responsible person (employer or manufacturer, etc.) to put right within a specified period of time any minor hazard or infringement of legislation.

7. **Issue a prohibition notice**. This requires the responsible person to stop immediately any activities that are likely to result in serious personal injury. This ban on activities continues until the situation is corrected. An appeal against an improvement or prohibition notice may be made to an industrial tribunal.

8. **Prosecute**. All persons, including employers, employees, self-employed, designers, manufacturers and suppliers who fail to comply with their safety duty may be prosecuted in a magistrates' court or in certain circumstances in the higher court system. Conviction can lead to unlimited fines, or a prison sentence, or both.

BBS Construction Services

STATEMENT OF COMPANY POLICY ON HEALTH AND SAFETY

The Directors accept that they have a legal and moral obligation to promote health and safety in the workplace and to ensure the cooperation of employees in this. This duty of care extends to all persons who may be affected by any operation under the control of BBS Construction Services.

Employees also have a statutory duty to safeguard themselves and others and to cooperate with management to secure a safe work environment.

The directors shall ensure, so far as reasonably practicable, that:
- Adequate resources and competent advice are made available in order that proper provision can be made for health and safety.
- Safe systems of working are devised and maintained.
- All employees are provided with all information, instructions, training and supervision required to secure the safety of all persons.
- All plant, machinery and equipment is safe and without risk to health.
- All places of work are maintained in a safe condition with safe means of access and egress.
- Arrangements are made for safe use, handling, storage and transport of all articles and substances.
- The working environment is maintained in a condition free of risks to health and safety and that adequate welfare facilities are provided.
- Assessment of all risks are made and control measures put in place to reduce or eliminate them.
- All arrangements are monitored and reviewed periodically.

These statements have been adopted by directors of the company and form the basis of our approach to health and safety matters.

Ivor Carpenter, Finance Director

Christine Whiteman, Human Resource Director

James Brett, Managing Director

The Director with responsibility for Health and Safety

Peter Brett, Chief Executive

Figure 2.7

BBS Construction Services

SITE SAFETY INDUCTION

As part of the company's commitment to safety the following checklist is provided for management, when employing new staff or on the transfer of existing staff to your site.

MANAGEMENT MUST:

- Issue and explain the company's Safety Policy.

- Introduce your site safety advisor/controller.

- Discuss and record any previous safety training and experience.

- Issue and discuss appropriate safety method statement.

- Emphasize the following points:
  – Emergency and first-aid procedures applicable to the site
  – Personal safety responsibilities, house keeping, hygiene and PPE
  – Need to report accidents, 'near misses' and unsafe conditions
  – Need for authorization and training, before use of all plant machinery and powered hand tools
  – Location of all welfare facilities

- Explain the procedure to be followed in the event of a health and safety dispute (consult safety advisor/controller in first instance).

- Show the site notice board (in the rest room) where safety notices and information are displayed.

- Finally inform of company's key phrase for all matters 'IF IN DOUBT ASK' then invite questions.

Figure 2.8

Health and safety

Chapter 2

Health and Safety Executive
Health and Safety at Work etc Act 1974, Sections 21, 23 and 24

Serial number
I

Improvement notice

Name
Address

Trading as*

Inspector's full name — I,
Inspector's official designation — one of Her Majesty's Inspectors of
Being an Inspector appointed by an instrument in writing made pursuant to section 19 of the said Act and entitled to issue this notice
Official address — of

Telephone number

hereby give you notice that I am of the opinion that
Location of premises or place of activity — at

you, as an employer/a self employed person/a person wholly or partly in control of the premises/other*
are contravening/have contravened in circumstances that make it likely that the contravention will continue or be repeated* the following statutory provisions:

SPECIMEN

The reasons for my said opinion are:

and I hereby require you to remedy the said contraventions or, as the case may be, the matters occasioning them by (and I direct that the measures specified in the Schedule which forms part of this Notice shall be taken to remedy the said contraventions or matters)*

Signature

An Improvement Notice is also being served on

of

related to the matters contained in this notice

Environment and Safety Information Act 1988 — This is a relevant notice for the purposes of the Environment and Safety
This page only will form the register entry*.

Signature Date

LP1 (rev 12/88) *See notes overleaf*

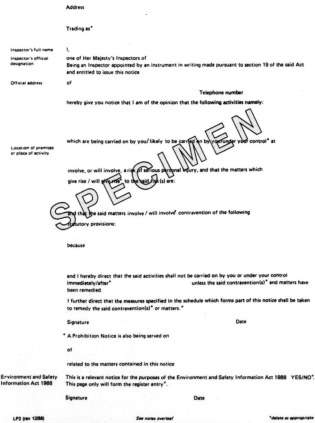

Figure 2.9

Specimen of an Improvement Notice (HSE)

Health and Safety Executive
Health and Safety at Work etc Act 1974, Sections 22, 23 and 24

Serial Number
P

Prohibition notice

Name
Address

Trading as*

Inspector's full name — I,
Inspector's official designation — one of Her Majesty's Inspectors of
Being an Inspector appointed by an instrument in writing made pursuant to section 19 of the said Act and entitled to issue this notice
Official address — of

Telephone number

hereby give you notice that I am of the opinion that the following activities namely:

which are being carried on by you/ likely to be carried on by you/under your control* at
Location of premises or place of activity

involve, or will involve, a risk of serious personal injury, and that the matters which give rise / will give rise* to the said risk (s) are:

SPECIMEN

and that the said matters involve / will involve* contravention of the following statutory provisions:

because

and I hereby direct that the said activities shall not be carried on by you or under your control immediately/after* unless the said contravention(s)* and matters have been remedied.

I further direct that the measures specified in the schedule which forms part of this notice shall be taken to remedy the said contravention(s)* or matters.*

Signature Date

* A Prohibition Notice is also being served on

of

related to the matters contained in this notice

Environment and Safety Information Act 1988 — This is a relevant notice for the purposes of the Environment and Safety Information Act 1988 YES/NO*.
This page only will form the register entry*.

Signature Date

LP2 (rev 12/88) *See notes overleaf* **delete as appropriate*

Figure 2.10

Specimen of a Prohibition Notice (HSE)

5. State **TWO** of the main objectives of the Health and Safety at Work Act.

6. State **TWO** duties of each of the following under the Health and Safety at Work Act:
   a) Employers
   b) Employees

7. State **THREE** main powers of a Health and Safety Executive Inspector.

8. Read the newspaper extract. State the reasons why the 22-year-old was fined £250.

BUILDING AND CONSTRUCTION
Harry Whiteman is Safety Consultant to BBS Contracts
SAFETY NEWS. June. PSB

Maidstone Crown Court fined a building contractor £25 000 recently over an incident where an 18-year-old trainee lost both hands and feet when the scaffold tube he was un-loading touched an overhead 33 000 volt electric cable. The Health and Safety Executive had asked in the magistrates court, where the maximum penalty is £2000, for this case to be referred to the Crown Court for sentence.

A major contractor in Birmingham city centre was recently fined £500 for supplying only two safety helmets for the 20 people who were employed on-site.

A 22-year-old site operative, who lost the sight in one eye as a result of a grinding wheel accident, was fined £250 in Northampton Magistrates' Court this week. The Health and Safety Executive who brought the prosecution claimed that the operative had failed to take notice of the safety sign or wear the safety goggles which had been supplied by the employer.

Figure 2.11

Management of health and safety legislation

safety tip

Always assess the risks and plan to take the appropriate action before undertaking any task.

Management of Health and Safety at Work Regulations (MHSWR)

These regulations apply to everyone at work. They require your employer and self-employed people to plan, control, organize, monitor and review their work. In doing this, they must:

◆ assess the risks associated with the work being undertaken
◆ have access to competent health and safety advice
◆ provide employees with health and safety information and training
◆ appoint competent persons in their workforce to assist them in complying with obligations under health and safety legislation
◆ make arrangements to deal with serious and imminently dangerous situations
◆ cooperate in all health and safety matters with others who share the workplace.

safety tip
Always wear your PPE.

As an employee, you are required to make full and proper use of anything put into place by your employer, to reduce the risk of injury during manual handling operations.

The Personal Protective Equipment at Work Regulations

Personal protective equipment (PPE) means all pieces of equipment, additions or accessories designed to be worn or held by a person at work to protect against one or more risks. Typical items of PPE are safety footwear, waterproof clothing, safety helmets, gloves, high visibility clothing, eye protection, dust masks, respirators and safety harnesses.

The use of PPE is seen as the **last** not the **first** resort. The first consideration is to undertake a risk assessment, with a view to preventing or controlling any risk at its source, by making machinery or work processes safer.

PPE requirements

◆ All items of PPE must be suitable for the purpose it is being used for and provision must be made for PPE maintenance, replacement and cleaning.
◆ Where more than one item is being worn, they must be compatible.
◆ Training must be provided in the correct use of PPE and its limitations.
◆ Employers must ensure that appropriate items are provided and are being properly used.
◆ Employees and the self-employed must make full use of PPE provided and in accordance with the training given. In addition any defect or loss must be reported to their employers.

The Health and Safety (Safety Signs and Signals) Regulations

Safety signs and signals legislation requires employers to provide safety signs in a variety of situations that do or could affect health and safety. There are four types of safety signs in general use. Each of these signs has a designated shape and colour, to ensure that health and safety information is presented to employees in a consistent, standard way, with the minimum use of words.

Details of these signs and typical examples of use are given in Table 2.1.

PROHIBITION — Stop/must not

WARNING — Risk of danger hazard ahead

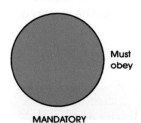

MANDATORY — Must obey

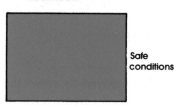

SAFE WAY TO GO — Safe conditions

Figure 2.14
Types of safety signs

Table 2.1 Safety signs

Purpose	Sign	Definition	Examples for use
Prohibition		A sign prohibiting certain behaviour	No smoking / Smoking and naked flames prohibited / Do not extinguish with water / Not drinking water / Pedestrians prohibited

Table 2.1 Safety signs continued

Purpose	Sign	Definition	Examples for use
Caution	⚠	A sign giving warning of certain hazards	Caution, risk of fire — Caution, toxic hazard — Caution, corrosive substance — General warning caution, risk of danger — Caution, risk of electric shock — Perimeter of hazard
Safe condition	▪	A sign providing information about safe conditions	First aid — Indication of direction — Indication of direction
Mandatory	●	A sign indicating that a special course of action is required	Head protection must be worn — Eye protection must be worn — Hearing protection must be worn — Foot protection must be worn — Hand protection must be worn — Respiratory protection must be worn
Supplementary	▭	A sign with text. Can be used in conjunction with a safety sign to provide additional information	IMPORTANT REPORT ALL ACCIDENTS IMMEDIATELY — SCAFFOLDING INCOMPLETE — SAFETY HELMETS ARE PROVIDED FOR YOUR SAFETY AND MUST BE WORN — PETROLEUM MIXURE HIGHLY FLAMMABLE NO SMOKING OR NAKED LIGHTS — WARNING HIGH VOLTAGE CABLES OVERHEAD — EYE WASH BOTTLE

In addition the following points, which are of particular concern to construction work, are highlighted in the regulations:

◆ In order to avoid confusion, too many signs should not be placed together
◆ Signs should be removed when the situation they refer to ceases to exist
◆ Fire-fighting equipment and its place of storage must be identified by being red in colour
◆ Traffic routes should be marked out using yellow
◆ Acoustic fire evacuation signals must be continuous and sufficiently loud to be heard above other noises on-site
◆ Anyone giving hand signals must wear distinctive, brightly coloured clothing and use the standard arm and hand movements.

Turn left   Turn right

Lower   Danger

Stop   Raise

Figure 2.15

Typical hand signals

For the following safety signs write on a separate piece of paper the type of sign and its meaning.

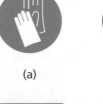

(a)   (b)   (c)   (d)

Figure 2.16   (e)   (f)   (g)   (h)

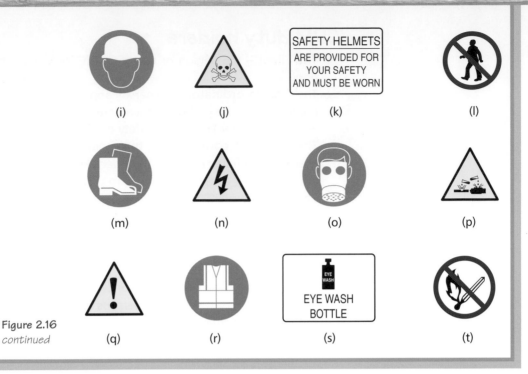

Figure 2.16
continued

Competent persons
can be defined as those
who have the experience,
knowledge and appropriate
qualifications, which enable
them to identify the risks
arising from a given situation
and the measures that are
required to mitigate the risks.

The Construction (Design and Management) Regulations (CDM)

Design and management legislation requires that health and safety is taken into account and managed during all stages of a construction project, from its conception, design and planning, throughout the actual construction process and afterwards during maintenance and repair.

These regulations apply to all construction projects irrespective of the size. The HSE must be notified where construction work is expected to:

◆ Last for more than 30 days, or
◆ Involve more than 500 person days of work, for example, 50 people working for over 10 days.

Health and safety plan and file

The regulations require the client, the designers and the building contractors to play their part in improving on-site health and safety. In doing this they will have to draw up a two-stage health and safety plan:

◆ *Phase* 1 – a pre-construction information pack, which highlights any particular risks of the project, and the equipment and the level of health and safety competence that a prospective contractor will require.

◆ *Phase* 2 – a construction phase health and safety plan, which sets out how health and safety will be managed during the project.

They will also have to draw up a health and safety file. This is to be produced during the project and passed on to the client or building user on completion. It should contain full operating details of the building and its systems, as well as details of health and safety risks that will have to be managed during future maintenance and cleaning operations.

Health and safety Chapter 2

CDM duty holders

◆ **The client** – the person or organisation, for whom the work is being done.
◆ **The CDM co-ordinator** – the person appointed by the client for notifiable projects, to advise and assist them with their duties and co-ordinate the arrangements for health and safety during all phases of the project.
◆ **The principal contractor** – the main contractor appointed by the client for notifiable projects, to undertake the construction work – may appoint other contractors and sub-contractors to undertake specific parts of the work.
◆ **The designer** – the person or persons who prepare drawings or specifications for a building including products to be used in a building and as such includes architects, engineers, surveyors and other product or component designers.
◆ **Contractors** – the main contractor on non-notifiable projects and sub-contractors and self-employed persons working on-site for all projects.
◆ **Workers and others** – everyone working on a building site or working for any of the above duty holders.

Client duties

◆ Ensure designers, contractors and all other team members are competent and have the necessary resources to undertake the project.
◆ Ensure they allow sufficient time and resources for each stage of the project from concept right through to completion.
◆ Ensure suitable management arrangements are in place throughout the project.
◆ Ensure contractors have made arrangements for suitable welfare facilities throughout the construction phase of the project.
◆ Ensure designers and contractors receive a pre-construction information pack.

Additional duties for notifiable construction projects:

◆ Appoint a CDM co-ordinator, who must be in place until the end of the construction phase.
◆ Provide the CDM co-ordinator with health and safety file information.
◆ Retain and provide access to the health and safety file on completion of the construction phase.
◆ Appoint a principal contractor, for the whole of the construction phase.
◆ Ensure the construction phase does not start until there are suitable welfare facilities and the construction phase plan is in place.

CDM co-ordinators duties

◆ Advise and assist the client in their duties under the CDM regulations.
◆ Inform the HSE where the construction project is deemed notifiable.
◆ Ensure good communication between client, designers and all contractors.
◆ Identify and pass on to all parties, a pre-construction information pack regarding the design of the structure and any foreseen health and safety issues.

◆ Work with the principal contractor regarding ongoing design and health and safety matters.
◆ Prepare and update the health and safety file.

Principal contractors duties

◆ Prepare, develop and implement a written health and safety plan and site rules, the initial plan must be completed before the actual construction phase begins.
◆ Plan, manage and monitor the construction phase in liaison with all contractors. All contractors should be given the relevant parts of the plan.
◆ Ensure suitable welfare facilities are provided from the start of a project and are maintained throughout the construction phase.
◆ Check the competence of all duty holders and ensure that all workers have site safety inductions and further information and training required for them to undertake their work in safety.
◆ Consult with all workers on health and safety issues.
◆ Work with the CDM co-ordinator regarding ongoing design and health and safety matters.
◆ Ensure that the site is secure at all times.

Contractor duties

◆ Plan, manage and monitor own work and that of their workers.
◆ Ensure that all of their appointees and workers are competent to undertake the work in hand.
◆ Provide workers with the required health and safety information and training.
◆ Ensure there are adequate welfare facilities for their workers.
◆ Comply with specific CDM health and safety construction site requirements.

Additional duties for notifiable construction projects:

◆ Check that the client is aware of their duties; a CDM co-ordinator has been appointed; and the HSE has been notified of the project before starting work.
◆ Co-operate with the principal contractor in the planning and managing of the construction work, including complying with reasonable directions and site rules.
◆ Provide the principal contractor with details of any sub-contractor they engage to carry out construction work.
◆ Provide any information needed for the health and safety file.
◆ Inform the principal contractor of problems with the health and safety plan or risks identified that may have implications on the project.
◆ Inform the principal contractor about reportable accidents, diseases and dangerous occurrences.

Workers and others duties

◆ Ensure that they are competent to undertake the work in hand.
◆ Co-operate with all duty holders and co-ordinate their own work so as to ensure the health and safety of construction workers and other persons who may be affected by the work.
◆ Report all obvious health and safety hazards and risks.

Requirements relating to health and safety on construction sites

The main objective is to promote the health and safety of employees, the self-employed and others who may be affected by construction activities and the ongoing maintenance of the building throughout its life.

The main specific requirements and 'good practices' are outlined as follows.

Competence

The client must take suitable steps to ensure that the appointed CDM coordinator, designers, principal contractor and other contractors are competent. No person should accept an appointment unless they are competent. No person should instruct a worker or manage design or construction work unless the worker is competent or under the supervision of a competent person.

Toilets (sanitary conveniences)

Suitable and sufficient sanitary conveniences (no specific number given) must be provided or made available at readily accessible places and as far as reasonably practicable should be adequately ventilated, lit and kept in a clean and orderly condition.

Separate rooms containing sanitary conveniences should be provided for men and women, with each convenience having a door that can be secured from the inside.

Washing facilities

Wash hand basins with hot and cold or warm water (should be running water where reasonably practicable) to be provided in the immediate vicinity of toilets and changing rooms. These must include soap and towels or other cleaning and drying facilities. Where the work is particularly dirty or involves exposure to toxic or corrosive substances, showers may be required. All rooms containing washing facilities must have adequate ventilation and lighting, and be kept in a clean, orderly condition. Unisex facilities are suitable for washing of hands, faces and arms, otherwise in a separate room that is used by one person at a time and can be locked from the inside.

Drinking water

Wholesome drinking water must be readily accessible in suitable places and clearly marked. Cups or other drinking containers must be provided, unless the water is supplied via a drinking fountain.

Storage and changing of clothing

Secure accommodation must be provided for normal clothing not worn at work and for protective clothing not taken home. Separate lockers may be required where there is a risk of protective clothing contaminating normal clothing. This accommodation should include changing facilities, seating and a means of drying wet clothing. Separate rooms, or use of rooms are required, where appropriate, for men and women to ensure privacy.

Rest facilities

Heated accommodation must be provided for taking breaks and meals. These facilities must include tables and chairs, a means of boiling water

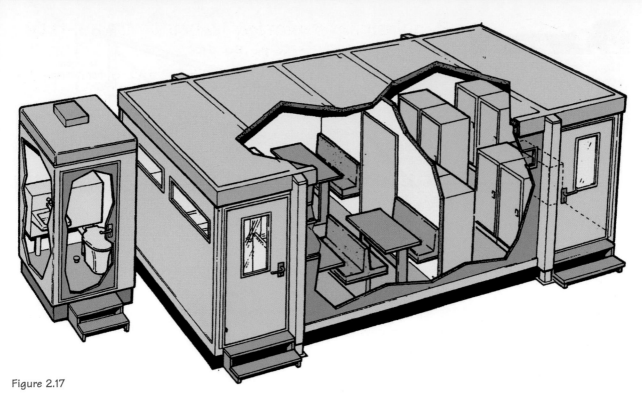

Figure 2.17

and a means of preparing food. Where necessary, they should include facilities for pregnant women and working mothers to rest lying down.

Prevention of falls

Edge protection is required to all working platforms and other exposed edges where it is possible to fall 2 metres or more. They should:

◆ Be sufficiently rigid for the purpose
◆ Include a guard-rail at least 950 mm above the edge
◆ Include a toe board at least 150 mm high
◆ Be subdivided with intermediate guard-rails, additional toe boards or brick guards etc., so that the maximum unprotected gap is 470 mm (Figure 2.18).

Other types of barrier may be used to protect edges, provided that they give the equivalent standard of protection against falls of persons and rolling or kicking of materials over the edge.

Support of excavations

Measures must be taken to prevent injury by collapsing excavations, falling materials or contact with buried underground services. Support for excavations is to be provided at an early stage. Sides of excavations must either be battered back to a safe angle or be supported with timbering or a proprietary system. All support work is to be carried out or altered by or under the supervision of a competent person. Measures must be taken to prevent people, materials or vehicles falling into excavations, for example: by the use of edge protection guard-rails; not storing materials, waste or plant items near excavations; keeping traffic routes clear of excavations (see Figure 2.21).

did you know?

Plant is the term used to describe all industrial machinery and vehicles used on a building site, such as cranes, excavators, earthmoving equipment, forklift and dumper trucks and power access equipment.

Health and safety Chapter 2

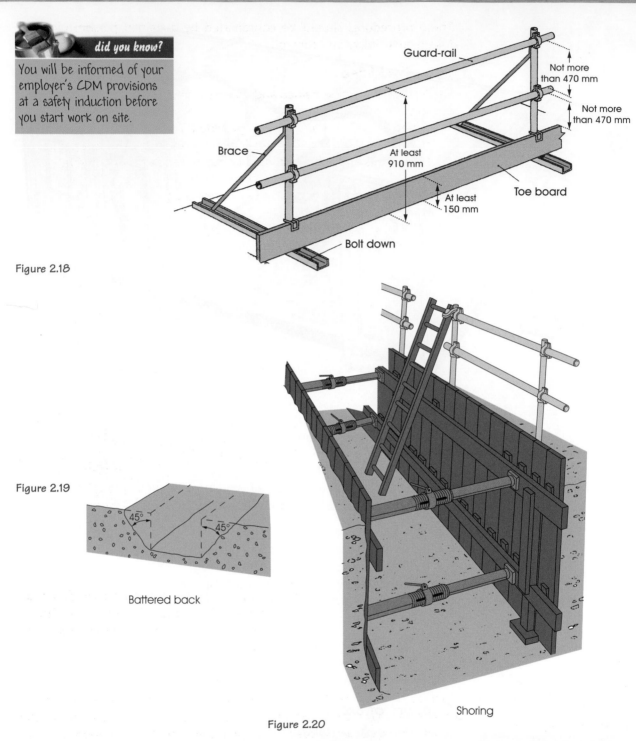

Figure 2.18

Guard-rail

Not more than 470 mm

Not more than 470 mm

Brace

At least 910 mm

Toe board

At least 150 mm

Bolt down

Figure 2.19

Battered back

45°   45°

Figure 2.20

Shoring

Emergency procedures

These are the arrangements made to deal with any unforeseen emergency, including fire, flooding, explosion and asphyxiation. These procedures must include the following:

◆ Provision of emergency signals, routes and exits for evacuation. These should be kept clear, be marked and illuminated
◆ Provision to notify the emergency services
◆ Provision of first aid and other facilities for treating and recovering injured persons.

These procedures should be coordinated by a trained person, who will take responsibility and control.

Use of vehicles

◆ All drivers must be trained and visiting drivers informed of site transport rules.
◆ Suitable traffic routes must be provided and clearly marked, avoiding sharp bends and blind corners including safe entry and exit points.
◆ Pedestrians and vehicles should be separated as far as possible.
◆ Reversing should be avoided wherever possible. Audible alarms are advisable where reversing is necessary.
◆ Provide trained signallers, wearing high visibility clothing, to assist drivers.

Figure 2.21

Fresh air

Every place of work or its approach, so far as reasonably practicable, should have sufficient fresh or purified air to ensure it is safe and without risks to health. Any plant used to provide this must give an audible or visual warning in the event of its failure.

Temperature and weather protection

Suitable and sufficient steps shall be taken to ensure so far as reasonably practicable that the temperature at any indoor place of work during working hours is reasonable (no specific temperature stated). Any outdoor place of work shall as far as reasonably practicable be so arranged or protective clothing provided in order to provide protection from adverse weather.

Lighting

Sufficient lighting shall be provided at every place of work, its approach and traffic routes; so far as reasonably practicable this should be by natural light. Where artificial lighting is used it should not affect the colour of health and safety signs. Emergency back-up lights should be provided for use in the event of the main artificial lighting system failing.

did you know?

A competent person must carry out inspections and make reports.

EXPLOSIVE

HIGHLY INFLAMMABLE

IRRITANT

CORROSIVE

TOXIC

HARMFUL

OXIDIZING

Figure 2.25

Control

Where the substance has to be worked with because either there is no choice, or because alternatives also present equal risks, exposure must be controlled by:

- Using the substance in a less hazardous form, e.g. use a sealed surface glass fibre insulation quilt rather than an open fibre one, to reduce the risk of skin contact or the inhalation of fine strands
- Using a less hazardous method of working with the substance, e.g. wet rubbing down of old lead-based painted surfaces rather than dry rubbing down, which causes hazardous dust; or applying spirit-based products by brush or roller rather than spraying
- Limiting the amount of substance used
- Limiting the amount of time people are exposed
- Keeping all containers closed when not in use
- Providing good ventilation to the work area. Mechanical ventilation may be required in confined spaces
- When cutting or grinding use tools fitted with exhaust ventilation or water suppression to control dust.

Study the table of recognized hazards in various construction jobs and suggested methods of control. See Table 2.2.

Protection

If exposure cannot be prevented or adequately controlled using any of the above, also use personal protective equipment (PPE).

- Always wear protective clothing, e.g. overalls, gloves (for protection and anti-vibration), boots, helmets, ear protection, eye protection goggles or visors and dust masks or respirators as appropriate.
- The use of barrier and after-work creams is recommended to protect skin from contact dermatitis.
- Ensure items of PPE are kept clean, so that they do not themselves become a source of contamination.
- All items of PPE should be regularly maintained, checked for damage and stored in clean, dry conditions.
- Replacement items of PPE and spare parts must be available for use when required.

Personal hygiene

Protection does not stop with PPE. Hazardous substance can be easily transferred from contaminated clothing and unwashed hands and face.

- Always wash hands and face before eating, drinking or smoking and also at the end of the working shift.
- Never eat, drink or smoke near to the site of exposure.
- Change out of contaminated work wear into normal clothes before travelling home.
- Have contaminated work wear regularly laundered (see Table 2.2).

Table 2.2 Hazardous substances in construction

Substances	Health risk	Jobs	Controls
DUSTS:			
Cement (Also when wet)	SK I ENT	Masonry, rendering	Prevent spread. Protective clothing, respirator when handling dry, washing facilities, barrier cream.
Gypsum	SK I ENT	Plastering	
Man-made mineral fibre	I SK ENT	Insulation	Minimize handling/cutting, respirator, one-piece overall, gloves, eye protection.
Silica	I	Sand blasting, grit blasting: scrabbling granite, polishing	Substitution – e.g. with grit, silica-free sand; wet methods; process enclosure/extraction; respirator.
Wood dust (Dust from treated timber, e.g. with pesticide, may present extra hazards)	I SK ENT	Power tool use in carpentry, especially sanding	Off-site preparation; on-site – enclosures with exhaust ventilation; portable tools – dust extraction; washing facilities; respirator.
Mixed dusts (Mineral and biological)	I SK ENT	Demolition and refurbishment	Minimize dust generation; use wet methods where possible; segregate or reduce number of workers exposed; protective clothing, respirator; good washing facilities/showers. Tetanus immunization.
FUMES/GASES:			
Various welding fumes from metals or rods	I	Welding/cutting activities	Mechanical ventilation in enclosed spaces; air supplied helmet; elsewhere good general ventilation. All work in confined spaces – exhaust and blower ventilation; self-contained breathing equipment confined space procedures. Position away from confined spaces. Where possible maintain exhaust filters; forced ventilation and extraction of fumes.
Hydrogen sulphide	I ENT	Sewers, drains, excavations, manholes	
Carbon monoxide/nitrous oxide	I	Plant exhausts	
SOLVENTS: In many construction products – paints, adhesives, strippers, thinners, etc.	I SK SW	Many trades, particularly painting, tile, fixing. Spray application is high risk. Most brush/roller work less risk. Regulation exposure increases risk	Breathing apparatus for spraying, particularly in enclosed spaces; use of mistless/airless methods. Otherwise ensure good general ventilation. Washing facilities, barrier cream.
RESIN SYSTEMS:			
Isocynates (MDI:TDI)	I ENT SK SW	Thermal insulation	Mechanical ventilation where necessary; respirators; protective clothing, washing facilities. Skin checks, respirator checks.
Polyurethane paints	I ENT SK SW	Decorative surface coatings	Spraying – airline/self-contained breathing apparatus; elsewhere good general ventilation. One-piece overall, gloves, washing facilities.
Epoxy	I SK SW	Strong adhesive applications	Good ventilation, personal protective equipment (respirator; clothing), washing facilities, barrier cream. As above.
Polyester	I SK ENT SW	Glass fibre claddings and coatings	
PESTICIDES: (e.g. timber preservatives, fungicides, weed killers)	I SK ENT SW	Particularly in-situ timber treatment. Handling treated timber	Use least toxic material. Mechanical ventilation, respirator, impervious gloves, one-piece overall and head cover. In confined spaces – breathing apparatus. Washing facilities, skin checks. If necessary biological checks. Handle only dry material.
ACIDS/ALKALIS:	SK ENT	Masonry cleaning	Use weakest solutions. Protective clothing, eye protections. Washing facilities (first aid including eye bath and copious water for splash removal).
MINERAL OIL:	SK I	Work near machines, compressers, etc. Mould release agents	Filters to reduce mist. Good ventilation. Protective clothing. Washing facilities; barrier creams. Skin checks.
SITE CONTAMINANTS: e.g. arsenic, phenols; heavy metals; micro-organisms etc. e.g. Wells syndrome, tetanus, hepatitis B	I SK SW	Site redevelopment of industrial premises or hospitals – particularly demolition ground work and drain/sewers	Thorough site examination and clearance procedures. Respirators, protective clothing. Washing facilities/showers. Immunization for tetanus.

Health risk
SK = skin; I = introduction; ENT = irritant eyes, nose, throat; SW = ingestion
Table extracted from Control of Substances Hazardous to Health (COSHH) Regulations

Health and safety Chapter 2

Monitoring and health surveillance

This must be carried out to ensure exposure to hazardous substances is being adequately controlled.

◆ Monitoring of the workplace is required to ensure exposure limits are not being exceeded, e.g. regular checks on noise levels and dust or vapour concentrations.

◆ Health surveillance is a legal duty for a limited range of work exposure situations (e.g. asbestos). However, many employers operate a health surveillance programme for all their employees. This gives medical staff the opportunity to check the general health of workers, as well as giving early indications of illness, disease and loss of sensory perception. Simple checks can be made on a regular basis including blood pressure, hearing, eyesight and lung peak flow. Any deterioration over time indicates the need for further action.

measuring up

9. Name **THREE** notices or certificates that must be displayed at a building site.

10. Explain what term 'reasonably practicable' means.

11. A safety sign that is contained in a yellow triangle with a black border is:
    a) prohibiting certain behaviour
    b) warning of certain hazards
    c) providing information about safety
    d) indicating that safety equipment must be worn.

12. Define what the term 'competent person' means.

13. Describe what is meant by 'work at height' and list the **THREE** measures that should be considered before working at height.

Information

Employers must provide their employees, who are or may be exposed to hazardous substances, with:

◆ Safety information and training for them to know the risks involved
◆ A safe working method statement, including any precautions to be taken or PPE to be worn
◆ Results of any monitoring and health surveillance checks.

safety tip

Always read safety method statements and take appropriate action.

Figure 2.26

BBS: Shopfitting Services
33 Stafford Thorne Street
Nottingham NG22 3RD
Tel. 0115 94000

SAFETY METHOD STATEMENT

Process:

The re-manufacture of MDF panel products. During this process a fine airborne dust is produced. This may cause skin, eye, nose and throat irritation. There is also a risk of explosion. The company has controls in place to minimize any risk. However, for your own safety and the protection of others, you must play your part by observing the following requirements.

General Requirements: At all times observe the following safety method statements and the training you have received from the company.
- Manual Handling
- Use of Woodworking Machines
- Use of Powered Hand Tools
- General House Keeping

Specific Requirements:
- When handling MDF, always wear gloves or barrier cream as appropriate. Barrier cream should be replenished after washing.
- When sawing, drilling, routing or sanding MDF, always use the dust extraction equipment and wear dust masks and eye protection.
- Always brush down and wash thoroughly to remove all dust, before eating, drinking, smoking, going to the toilet and finally at the end of the shift.
- Do not smoke outside the designated areas.
- If you suffer from skin irritation or other personal discomfort seek first-aid treatment or consult the nurse.

IF IN DOUBT ASK

Hazards associated with noise

Exposure to loud noise can permanently damage the hearing resulting in deafness or tinnitus (ringing in the ear).

Under the Noise at Work Regulations certain duties are placed on manufacturers, suppliers, employers and employees to reduce the risk of hearing damage to the lowest reasonably practicable level.

◆ *Manufacturers/suppliers* – it is a requirement to provide low-noise machinery and to provide information to purchasers concerning the level of noise likely to be generated and its potential hazard.

◆ *Employers* – it is a requirement to reduce the risk of hearing damage. Specific actions are triggered at daily personal noise exposure levels of 80 and 85 dB(A) and peak sound pressures of 135 and 137 dB(C). The maximum exposure limits are 87 dB(A) and 140 dB(C). At 80 dB(A) employers must:

  ► make a noise assessment to identify workers exposed and the actions to be taken
  ► provide suitable ear protectors on request
  ► provide information to employees about the risks to hearing and the legislation.

did you know?

Noise is measured in decibels (dB):
- An A-weighting dB(A) is used to measure average noise levels
- A C-weighting dB(C) is used to measure peak levels.

Typical examples of noise levels on a construction site are:
- Conversation 50–60 dB(A)
- Dumper truck 75–85 dB(A)
- Power drill 80–90 dB(A)
- Concrete breaker 90–100 dB(A)
- Chainsaw 100–110 dB(A)

Health and safety

Chapter 2

At 85 dB(A) employers must:

> ▶ reduce noise exposure as far as is reasonably practicable by means other than hearing protectors, e.g. by erecting acoustic enclosures or barriers around machines or noise producing operations, changing to quieter tools, fitting silencers or changing patterns of work to ensure combinations of the noisier machines are not used at the same time;
> ▶ designate the work area as an ear protection zone, by suitable signs indicating 'Ear protection must be worn';
> ▶ provide suitable ear protectors and ensure that all who enter the area wear them.

◆ *Employees* – have the duty to use the noise control equipment and hearing protection that is provided and to report any defects.

Hazards associated with vibration

Exposure to vibration can cause permanent damage to anyone who operates a machine or power tool. Effects include impaired blood circulation, damage to nerves and muscles and damage to bones in the hands and arms.

Hand–arm vibration syndrome (HAVS)

This is likely in any process where hands are exposed to vibrations from vibrating machines, tools or work pieces. The effect of the vibration dose received by an operator over a day depends on:

◆ The vibration frequency
◆ The duration of exposure
◆ The exposure pattern
◆ The grip and force required in guiding the tool or work piece.

Precautions should be taken where:

◆ Any tingling or numbness is felt after 5–10 minutes of use
◆ High-risk operations are carried out, such as the use of hand-held sanders, hand-fed or hand-held circular saws, pneumatic nailing or stapling tools.

◆ *Employers* – should take preventive measures to reduce vibration, such as limiting the duration of exposure, the selection of suitable work methods, use of reduced vibration tools and appropriate PPE. In addition, health surveillance of people who are exposed to vibration should also be undertaken, especially where high levels or long durations are concerned.

◆ *Manufacturers* – have a duty to reduce vibration levels and supply vibration data for their equipment. There is a 'traffic light' labelling system for use particularly on power tools, which gives the operator guidance on vibration levels and recommended daily maximum duration of use.

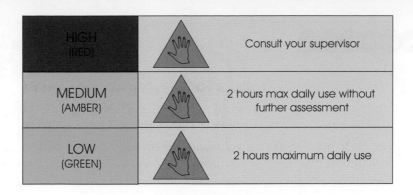

HIGH (RED)	⚠	Consult your supervisor
MEDIUM (AMBER)	⚠	2 hours max daily use without further assessment
LOW (GREEN)	⚠	2 hours maximum daily use

Figure 2.27

General site safety

It should be the aim of everyone to prevent accidents. Remember, you are required by law to be aware and fulfil your duties under the Health and Safety at Work Act and other regulations.

The main contribution you as an operative can make towards the prevention of accidents is to work in the safest possible manner at all times, thus ensuring that your actions do not put at risk yourself, your workmates or the general public.

Safety: on-site and in the workshop

A safe working area is a tidy working area. All unnecessary obstructions, which may create a hazard, should be removed, e.g. offcuts and unwanted materials, disused items of plant, and the extraction or flattening of nails from discarded pieces of timber. Therefore:

safety tip

Keep fire extinguishers accessible.

◆ Clean up your work bench/work area periodically, as offcuts and shavings are potential tripping and fire hazards
◆ Learn how to identify the different types of fire extinguishers and what type of fire they can safely be used on. Staff in each work area should be trained in the use of fire extinguishers.
◆ Careful disposal of waste from heights is essential. Any waste should always be lowered safely by hoist or in an enclosed chute and not thrown or dropped from scaffolds and window openings, etc. Even a small bolt or fitting dropped from a height can penetrate a person's skull and almost certainly lead to brain damage or death.

Figure 2.28 *Control of these three elements, or the removal of any one of them, will prevent fires from starting*

Health and safety

Chapter 2

safety tip

Fire extinguishers can give off dangerous fumes.

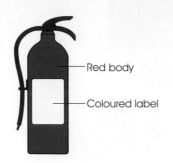

Red body

Coloured label

Fire Classification

Class A:
Wood, paper, textiles and any other carbonaceous materials

Class B:
Flammable gases such as petrol, oils, fats and paints

Class C:
Flammable gases such as propane, butane and natural gas

safety tip

Always stack materials safely.

TYPE OF FIRE RISK	USE OF FIRE EXTINGUISHERS				
	White label	Cream label	Black label	Blue label	Red
	Water	Foam	Carbon dioxide	Dry powder	Fire blanket
Class A	✓	✓	✗	✓	Can be used for smothering all types of fire. Also for use where clothing is alight since it does not pose a risk to skin or to breathing as some extinguishers do
Class B	✗	✓	✓	✓	
Class C	✗	✗	✓	✓	
Electrical	✗	✗	✓	✓	
Vehicle	✗	✓	✗	✓	

Figure 2.29

did you know?

Three essential features are needed for a fire to start:
- **Fuel** – any combustible or flammable material, which can be a solid, liquid or a gas
- **Oxygen** – normally from the air, but can also be given off by certain chemicals
- **Heat or ignition source** – such as naked flames, cigarette ends, electric sparks, overheating equipment, welding or burning activities etc.

These make up the three elements of the fire triangle.

◆ Ensure your tools are in good condition. Blunt cutting tools, loose hammer heads, broken or missing handles and mushroom heads must be repaired immediately or the use of the tool discontinued.

◆ When moving materials and equipment, always look at the job first; if it is too big for you then get help. Look out for splinters, nails and sharp or jagged edges on the items to be moved. Always lift with your back straight, elbows tucked in, knees bent and feet slightly apart. When putting an item down ensure that your hands and fingers will not be trapped.

◆ Materials must be stacked on a firm foundation; stacks should be of reasonable height so as to allow easy removal of items. They should also be bonded to prevent collapse and battered to spread the load. Pipes and drums etc. should be wedged or chocked to prevent rolling. Never climb on a stack or remove material from its sides or bottom.

◆ Excavations and inspection chambers should be either protected by a barrier or covered over completely to prevent people carelessly falling into them.

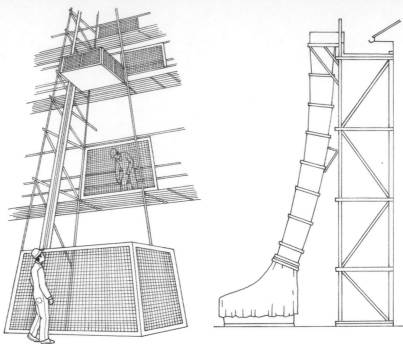

Figure 2.30

Correct lowering of waste from heights

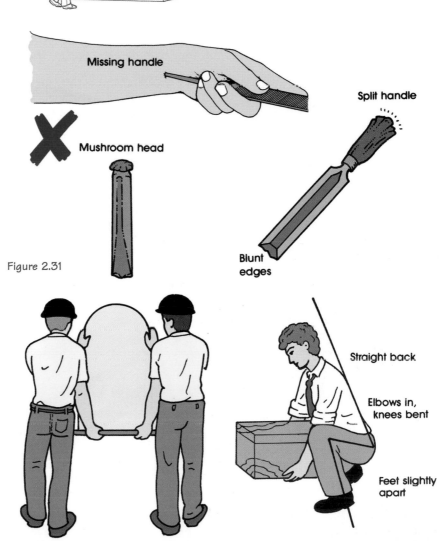

Missing handle

Split handle

X Mushroom head

Blunt edges

Figure 2.31

Straight back

Elbows in, knees bent

Feet slightly apart

Figure 2.32 *Paired lifting*

Figure 2.33 *Correct lifting position*

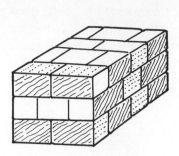

Figure 2.34 *Bonded material storage*

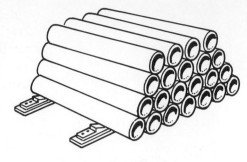

Figure 2.35 *Chocked material storage*

◆ Extra care is needed when working at heights. Ladders should be of sufficient length for the work in hand and should be in good condition and not split, twisted or with rungs missing. They should also be used at a working angle of 75 degrees and securely tied at the top. This angle is a slope of four vertical units to one horizontal unit. Where a fixing at the top is not possible, an alternative is the stake-and-guy rope method illustrated in Figure 2.37. Otherwise, arrange for someone to stand on the foot of the ladder. (The 'footer' must wear the appropriate headgear and pay attention to the task at all times and not 'watch the scenery'.) Wooden ladders must not be painted as this may hide defects. Ensure that extension ladders have sufficient

Figure 2.36

Protection of excavations

overlap for strength (at least two rungs for short ladders and up to four for longer ones) and that the latching hook is engaged. Never overreach when working on a ladder: always take the time to stop and move it closer to the work position. Ladders should be lowered and locked away at night.

Figure 2.37

Safe working angle and security of ladders

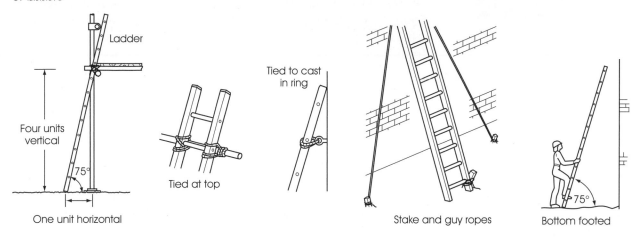

◆ Scaffolds should be inspected before working on them. Check to see that all components are there and in good condition, not bent, twisted, rusty, split, loose or out of plumb, and are level. Also ensure that the base has not been undermined or is too close to excavations. If in doubt do not use, and have it looked at by an experienced scaffolder or report it to your supervisor. Never remove any part from a scaffold: you may be responsible for its total collapse. Never block a

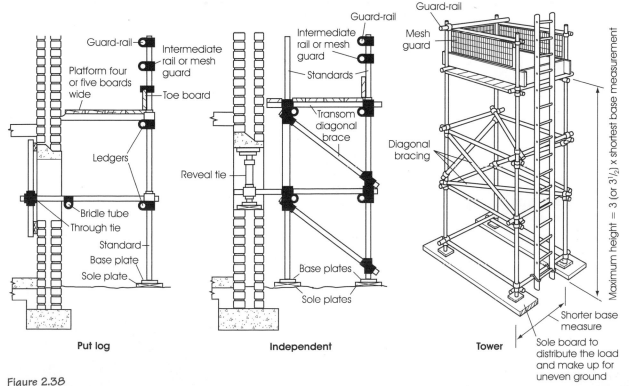

Figure 2.38

Types of scaffold with safety features

Health and safety

Chapter 2

scaffold with your tools, equipment or materials and always clear up any mess made as you go (don't leave it to form a hazard).

◆ You may be required to erect and use other working platforms, hop-ups, split-head type working platforms, stepladders and trestle scaffolds, during the execution of your work. This type of equipment is only generally suitable for internal use and at restricted heights. Ensure all the equipment is in good order and that it is only erected on a flat level surface. Never overreach when working on ladders, steps or trestles and ensure your knees are always below the top of the steps.

safety tip

Always check working platforms before using them.

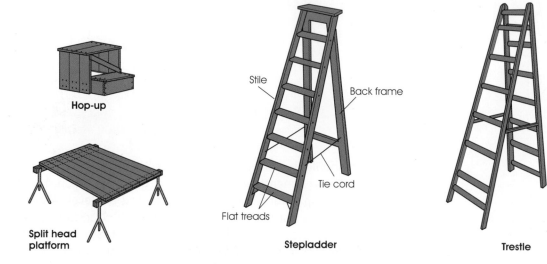

Hop-up

Split head platform

Stile

Back frame

Tie cord

Flat treads

Stepladder

Trestle

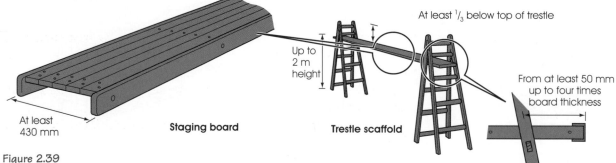

At least ¹/₃ below top of trestle

Up to 2 m height

At least 430 mm

Staging board

Trestle scaffold

From at least 50 mm up to four times board thickness

Figure 2.39

Other working platforms

safety tip

Always use the correct equipment for the work in hand and ensure it is safe.

◆ When working on roofs, roofing ladders or crawl boards should be used to provide safe access and/or to avoid falling through fragile coverings.

◆ Working with electrical and compressed-air equipment brings additional hazards as they are both potential killers. Installations and equipment should be checked regularly by qualified personnel; if anything is incomplete, damaged, frayed, worn or loose, do not use it, but return it to stores for attention. Ensure cables and hoses are kept as short as possible and routed safely out of the way to prevent risk of tripping and damage, or in the case of electric cables, from lying in damp conditions.

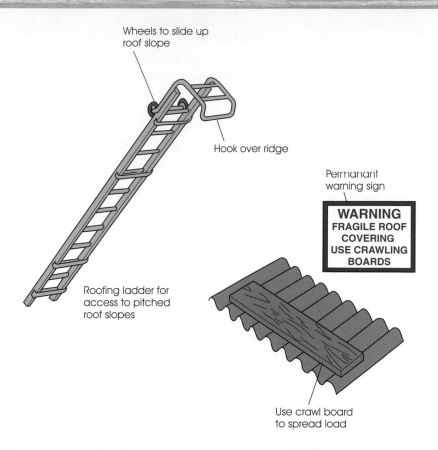

Wheels to slide up
roof slope

Hook over ridge

Permanant
warning sign

WARNING
FRAGILE ROOF
COVERING
USE CRAWLING
BOARDS

Roofing ladder for
access to pitched
roof slopes

Use crawl board
to spread load

Figure 2.40

Safe access to roofs

Protective clothing and equipment

Always wear the correct personal protective equipment (PPE) for the work
in hand:

◆ Safety helmets, safety footwear and a high visibility vest should be
worn at all times
◆ Wear ear protectors when carrying out noisy activities, and safety
goggles when carrying out any operation that is likely to produce dust,
chips or sparks, etc.
◆ Dust masks or respirators should be worn where dust is being produced
or fumes are present
◆ Wear gloves when handling materials
◆ Wet weather clothing is necessary for inclement conditions.

Many of these items must be supplied free of charge by employers.

safety tip

Always wear the appropriate
protection for the work in
hand.

Figure 2.41

Typical range of PPE

Health and safety

Chapter 2

Personal hygiene

Care should be taken with personal hygiene, which is just as important as physical protection. Some building materials have an irritant effect on contact with the skin. Some are poisonous if swallowed, while others can result in a state of unconsciousness (narcosis) if their vapour or powder is inhaled. These harmful effects can be avoided by wearing the appropriate PEE and by taking proper hygiene precautions:

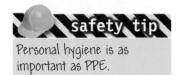

Personal hygiene is as important as PPE.

◆ Follow the manufacturer's instructions
◆ Look at the warning signs that are displayed on packaging
◆ Avoid inhaling fumes or powders
◆ Wear a barrier cream
◆ Thoroughly wash your hands before eating, drinking, smoking and after work.

HIGHLY FLAMMABLE (F)

HARMFUL (X)

TOXIC (T)
VERY TOXIC (T+)

OXIDISING (O)

EXTREMELY FLAMMABLE (F+)

DANGEROUS FOR THE ENVIRONMENT (N)

CORROSIVE (C)

IRRITANT (XI)

Figure 2.42

Examples of warning signs on packaging

First aid

First aid is the treatment of persons with the purpose of preserving life until medical help is obtained and also the treatment of minor injuries for which no medical help is required.

In all cases only a trained first-aider should administer first aid. Take care not to become a casualty yourself. Send for the nearest first-aider and/or medical assistance (phone 999) immediately. A record should be made of all accidents and first aid treatments.

Even minor injuries where you may have applied a simple plaster or sterilized dressing could become infected and require further attention. You are strongly recommended to seek medical attention if a minor injury becomes inflamed, painful or festered.

It is recommended that you read the first-aid guidance leaflet, which should be found in every first aid box.

Figure 2.43

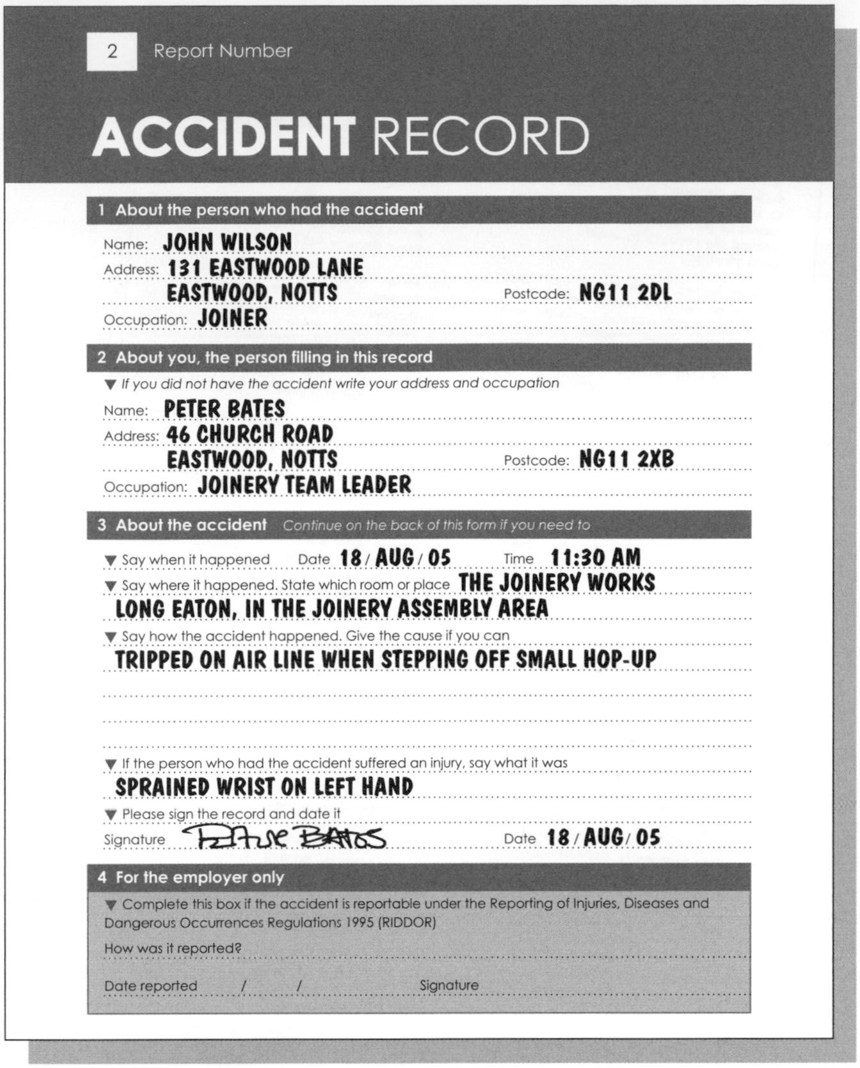

2 Report Number

ACCIDENT RECORD

1 About the person who had the accident

Name: **JOHN WILSON**
Address: **131 EASTWOOD LANE**
EASTWOOD, NOTTS   Postcode: **NG11 2DL**
Occupation: **JOINER**

2 About you, the person filling in this record

▼ If you did not have the accident write your address and occupation

Name: **PETER BATES**
Address: **46 CHURCH ROAD**
EASTWOOD, NOTTS   Postcode: **NG11 2XB**
Occupation: **JOINERY TEAM LEADER**

3 About the accident   Continue on the back of this form if you need to

▼ Say when it happened   Date **18 / AUG / 05**   Time **11:30 AM**
▼ Say where it happened. State which room or place **THE JOINERY WORKS**
LONG EATON, IN THE JOINERY ASSEMBLY AREA
▼ Say how the accident happened. Give the cause if you can
TRIPPED ON AIR LINE WHEN STEPPING OFF SMALL HOP-UP

▼ If the person who had the accident suffered an injury, say what it was
SPRAINED WRIST ON LEFT HAND
▼ Please sign the record and date it
Signature *Peter Bates*   Date **18 / AUG / 05**

4 For the employer only

▼ Complete this box if the accident is reportable under the Reporting of Injuries, Diseases and Dangerous Occurrences Regulations 1995 (RIDDOR)
How was it reported?

Date reported   /   /   Signature

Figure 2.44

Health and safety testing in construction

In order to raise safety standards in construction, employers and their clients are demanding that all workers and visitors to sites have current membership of a competence-based health and safety at work registration scheme to construction sites.

The Construction Skills Certification Scheme (CSCS), which is administered by the Construction Industry Training Board (CITB), has the greatest uptake of all recognized schemes.

CSCS cards provide evidence that the cardholders are competent and have up-to-date knowledge of health and safety legislation concerning construction site activities associated with their work.

To obtain a CSCS card you will have to attend a Test Centre and answer 35 to 40 random multiple-choice questions taken from a substantial bank relating to all aspects of health and safety in construction.

(b) *Chemical burns* Remove any contaminated clothing which shows no sign of sticking to the skin and flush all affected parts of the body with plenty of clean, cool water ensuring that all the chemical is so diluted as to be rendered harmless. Apply a sterilized dressing to exposed, damaged skin and clean towels to damaged areas where the clothing cannot be removed. (N.B. Take care when treating the casualty to avoid contamination.)

(c) *Foreign bodies in the eye* If the object cannot be removed readily with a clean piece of moist material, irrigate with clean, cool water. People with eye injuries which are more than minimal must be sent to hospital with the eye covered with an eye pad from the container.

(d) *Chemical in the eye* Flush the open eye at once with clean, cool water; continue for at least 5 to 10 minutes and, in any case of doubt, even longer. If the contamination is more than minimal, send the casualty to hospital.

(e) *Electric shock* Ensure that the current is switched off. If this is impossible, free the person, using heavy duty insulating gloves (to BS 697/1977) where these are provided for this purpose near the first-aid container, or using something made of rubber, dry cloth or wood or a folded newspaper; use the casualty's own clothing if dry. *Be careful* not to touch the casualty's skin before the current is switched off. If breathing is failing or has stopped, start resuscitation and continue until breathing is restored or medical, nursing or ambulance personnel take over.

(f) *Gassing* Move the casualty to fresh air but *make sure that whoever does this is wearing suitable respiratory protection.* If breathing has stopped, start resuscitation and continue until breathing is restored or until medical, nursing or ambulance personnel take over. If the casualty needs to go to hospital make sure a note of the gas involved is sent with him.

General

(a) *Hygiene* When possible, wash your hands before treating wounds, burns or eye injuries. Take care in any event not to contaminate the surfaces of dressings

(b) *Treatment position* Casualties should be seated or lying down while being treated

(c) *Record-keeping* An entry must be made in the accident book (for example B1 510 Social Security Act Book) of each case

(d) *Minor injuries* Casualties with minor injuries, of a sort they would attend to themselves if at home, may wash their hands and apply a small sterilized dressing from the container

(e) *First aid materials* Each article used from the container should be replaced as soon as possible

Health and Safety (First Aid) Regulations 1981

General first aid guidance for first aid boxes

Note: Take care not to become a casualty yourself while administering first aid. Be sure to use protective clothing and equipment where necessary. If you are not a trained first-aider, send immediately for the nearest first-aider where one is available.

Advice on treatment

If the assistance of medical or nursing personnel will be required, send for a doctor or nurse (where they are employed at the workplace) or ambulance immediately. When an ambulance is called, arrangements should be made for it to be directed to the scene without delay.

Priorities

(1) *Breathing* If the casualty has stopped breathing, resuscitation must be started at once *before any other treatment is given* and should be continued until breathing is restored until medical, nursing or ambulance personnel take over.

Mouth-to-mouth resuscitation

(2) *Bleeding* If bleeding is more than minimal, control it by direct pressure – apply a pad of sterilized dressing or, if necessary, direct pressure with fingers or thumb on the bleeding point. Raising a limb if the bleeding is sited there will help reduce the flow of blood (unless the limb is fractured).

(3) *Unconsciousness* Where the patient is unconscious, care must be taken to keep the airway open. This may be done by clearing the mouth and ensuring that the tongue does not block the back of the throat. Where possible, the casualty should be placed in the recovery position.

Recovery position

(4) *Broken bones* Unless the casualty is in a position which exposes him to further danger, do not attempt to move a casualty with suspected broken bones or injured joints until the injured parts have been supported. Secure so that the injured parts cannot move.

(5) *Other injuries*

(a) *Burns and scalds* Small burns and scalds should be treated by flushing the affected area with plenty of clean cool water before applying a sterilized dressing or a clean towel. Where the burn is large or deep, simply apply a dry sterile dressing. (N.B. Do not burst blisters or remove clothing sticking to the burns or scalds.)

Figure 2.45

CONSTRUCTION SKILLS CERTIFICATION SCHEME

CSCS

MR P S BRETT

tion No:00

Date: End Jul 20

CONSTRUCTION SITE VISITOR

The authenticity of this card can be checked by telephoning 01405 57877

Registration No: 00

Employers & Site Managers remain responsible under the Health & Safety at Work Act for ensuring that all holders of this card receive site induction and are employed as appropriate.

The cardholder has met the Health and Safety Awareness requirements as laid out in the CSCS Scheme Booklet

CSCS is a registered Certification Mark

EXAMPLE

Figure 2.46

Site security

Security controls on construction sites are concerned with the loss or theft of tools, equipment and materials as well as the protection of the general public.

Construction sites can be fairly easy targets for would-be thieves. High staff turnover and the high value of materials, plant and equipment can result in a quick profit for successful thieves. This is coupled with the fact that people working on-site commit a proportion of theft from construction sites.

The following measures can be used to raise awareness of security issues:

◆ All members of the workforce should be made personally responsible for company equipment that they use. If work equipment that you are using is lost through your carelessness or negligence you could be subject to disciplinary action
◆ All members of the workforce should be made aware of the company crime policy and be familiar with site security procedures
◆ Everyone on-site should be made responsible for reporting suspicious incidents to management
◆ All theft of materials, plant and equipment should be reported immediately by management to the police.

Perimeter protection

Fences and hoardings are the best means of preventing unauthorized access to a construction site. Entrances and exits should be kept to a minimum. The reception area for site visitors and delivery drivers should be located at the main gate and may be operated by a security guard on larger sites. Other perimeter protection considerations include:

◆ Display of warning notices around perimeter, which state that '**Security precautions are in force around the perimeter of this site**'
◆ CCTV cameras and alarm systems to protect the perimeter, compounds, stores and offices etc.
◆ Light the perimeter and access gates out of working hours; some of these may be operated by sensors.

Compounds and stores

Large valuable items should be stored in a fully fenced security compound within the perimeter hoarding; smaller valuable items should be kept in a secure, lockable store under the control of a store person, who is responsible for their issue to authorized personnel only. Stores should have a minimum of windows, which must be protected from 'break-in' by mesh grills or lockable shutters.

Security marking

To discourage theft all company property, plant and equipment can be permanently marked, using either paint, engraving or ultraviolet markings.

Security notice to all workers

The following notice can be displayed in the site changing room or canteen to advise people of their responsibilities with regards to site security:

BBS: Construction Site Security

The company aim to make this a secure site: we need your help in keeping it that way.

You have a duty to report any suspicious behaviour to your supervisor or the site manager. All reports willl be investigated and information given will be treated in the strictest confidence.

The company has the policy of: **Always Prosecuting Thieves**.

Theft and vandalism on our construction sites loses money and could result in job losses. It's in all our interests to observe the following rules:

- Ensure all tools are locked up when not in use
- Ensure all property is clearly marked
- Ensure ignition keys are removed from unattended plant and vehicles
- Ensure all company vehicles are locked and parked up within the site perimeter overnight and at weekends
- Ensure all keys are returned to the site manager.

Remember: If it's **on-site**, make sure it's either **in-sight** or securely locked up and **safe**.

Figure 2.47

measuring up

14. The correct angle for an access ladder to a scaffold is:
    a) 1 unit horizontal, 2 units vertical
    b) 1 unit horizontal, 4 units vertical
    c) 2 units horizontal, 1 unit vertical
    d) 4 units horizontal, 1 unit vertical

15. State **FOUR** building site operations where you would insist on the use of protective equipment. Name the item of protective equipment in each case.

16. Describe **FOUR** general procedures to be followed, which would aid either site or workshop safety.

17. Produce a list to explain **FOUR** responsibilities that site operatives have concerning site security.

18. Name the **THREE** elements that are required for a fire to start.

19. Explain why painted wooden ladders should not be used.

20. State what the abbreviation HAVS means.

21. Name the contents of a red fire extinguisher having a white label and state the type of fires it is safe to use it for.

activity

Hazard spotting

The following is intended to reinforce the work undertaken in this book. It gives you an opportunity to use your newly acquired safety awareness.

Figure 2.48

From the illustration of an unsafe building site in Figure 2.49 (overleaf), you are required to identify safety hazards, breaches of regulations and general bad practices, etc.

There are at least 20 to be found, which relate to areas covered in this chapter.

How many hazards can you spot?

Photocopy Figure 2.49 (overleaf), then circle each hazard etc. and number it as in the sketch on the left:

Then describe each hazard, etc. like this:

1. Sole plate missing from under scaffold standard.

Figure 2.49
An unsafe building site

Communications

This chapter is intended to provide the new entrant with an overview of the various means of communication used in the construction industry. Its contents are assessed in the **NVQ Unit No. VR 02 Conform to Efficient Work Practices**. It is concerned with verbal, written and visual communication used in the workplace on a day-to-day basis.

In this chapter you will cover the following range of topics:

◆ Methods of communication
◆ Principles of construction drawings
◆ Construction activity documents
◆ Building control
◆ The construction process and programme
◆ Site and workshop communication
◆ Employment rights and documentation
◆ Personal communications.

Methods of communication

There are three main methods of communication used in the construction industry: speech; written or printed information; and visual and graphic information.

Speech

Speech refers mainly to people talking and listening, but also gives the opportunity of observing and interpreting facial expressions and other body language.

On-site this may include the following: general conversations; giving or receiving briefings and instructions; giving or receiving training; taking or making telephone calls; participation at site meetings; participation in interview situations etc.

More time is spent listening and talking than any other forms of communication. It gives us direct personal contact; the opportunities to ask and answer questions; seek clarifications; discuss and resolve misunderstandings as they arise. However there are a number of disadvantages:

- Disputes can arise at a later stage, as there are no written records
- May not always be enough time to think clearly or interpret what has been said and respond accordingly
- May not be the most effective where large numbers of people are involved
- Often harder for an individual to voice comment or criticism.

Written or printed information

This refers mainly to people either writing new material or reading material produced by others.

On-site this may include the writing or reading of the following: orders; records; reports; forms; letters, memos; e-mails; faxes; notices; posters; minutes of meetings; specifications; descriptions and bills of quantities etc.

Written communication has a number of advantages over other methods of communication:

- Can be used to provide confirmation of earlier verbal communications
- Complicated ideas and details can often be more easily understood
- Can be used to provide evidence of actions taken or tasks completed
- Can be referred to at a later stage in the event of a dispute or query arising.

Visual and graphic information

This refers mainly to people using or producing drawings, pictures and symbols.

On-site this may include the interpretation or production of plans; drawings; details; sketches; diagrams; tables; graphs; charts; photographs, video and architectural models etc.

Visual communication is considered very effective for the following reasons:

- Complicated ideas and details are often easier to understand visually, rather than the same information given via the spoken or written form
- Clearly shows the whole process, details, methods and techniques
- Illustrations and visual aids help to aid understanding, when used to support spoken or written information
- Aids understanding for those with reading or language difficulties.

Principles of construction drawings

Drawings are the major means used to communicate technical information between all parties involved in the building process. They must be clear, accurate and easily understood by everyone who uses them. In order to achieve this, it is essential that architects, designers and all others who produce drawings use standardized methods for scale, line-work, layout, symbols and abbreviations.

Communications Chapter 3

Scales and lines used on drawings

Scales use ratios that permit measurements on a drawing or model to relate to the real dimensions of the actual job. It is impractical to draw buildings, plots of land and most parts of a building to their full size, as they simply will not fit on a sheet of paper. Instead they are normally drawn to a smaller size, which has a known scale or ratio to the real thing. These are then called scale drawings, as illustrated in Figure 3.1.

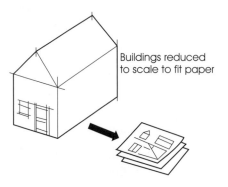

Buildings reduced to scale to fit paper

Figure 3.1

Scale drawings

The main scales (ratios) used in the construction industry are:

1:1; 1:5; 1:10; 1:20; 1:50; 1:100; 1:200; 1:500; 1:1250; 1:2500

We say: '1:10' is a ratio or scale of 'one to ten' or sometimes 'one in ten'.

The ratio shows how much smaller the plan or model is to the original. A 1-metre length drawn to various scales is illustrated.

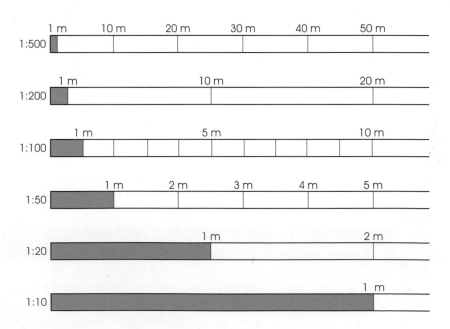

Figure 3.2

1 metre drawn to various scales

In a drawing to a scale of 1:10:

◆ 5 mm would stand for or relate to 50 mm on the actual job; or
◆ 50 mm would stand for or relate to 0.5 metre (500 mm).

In a drawing to a scale of 1:20:

◆ 5 mm would stand for or relate to 100 mm on the actual job; or
◆ 50 mm would stand for or relate to 1 metre (1000 mm).

It is simply a matter of multiplying the scale measurement by the scale ratio to determine the actual size to which it relates.

example

50 mm at a 1:10 scale ratio represents 500 mm in the building. 50 mm at a 1:20 scale ratio represents 1000 mm in the building.

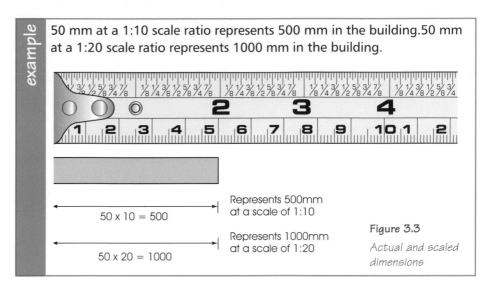

50 x 10 = 500   Represents 500mm at a scale of 1:10

50 x 20 = 1000   Represents 1000mm at a scale of 1:20

Figure 3.3

Actual and scaled dimensions

activity

Complete the table below.

Scale ratio of job	Size drawn	Actual size
1:1	100 mm	100 mm
1:5	250 mm	1250 mm
1:10	100 mm	1000 mm
1:20	75 mm	
1:50	125 mm	
1:100	150 mm	
1:200	125 mm	
1:1500	45 mm	
1:1250	25 mm	
1:2500	50 mm	

Communications *Chapter 3*

Scale rules

These are used to save having to calculate the actual size represented on drawings. They have a series of scales on them so that actual dimensions can be read directly from the rule.

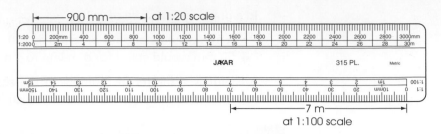

Figure 3.4

Scale rule

Although scale rules can be useful when reading drawings, preference should always be given to written dimensions shown on a drawing. Mistakes can be made due to the dimensional instability of paper at differing moisture contents or where a fold in the paper might make it impossible to get an accurate size.

Select a scale rule and mark on it lines representing the following:

- 7 metres to a scale of 1:50
- 1200 mm to a scale of 1:100
- 600 mm to a scale of 1:200
- 85 metres to a scale of 1:1250.

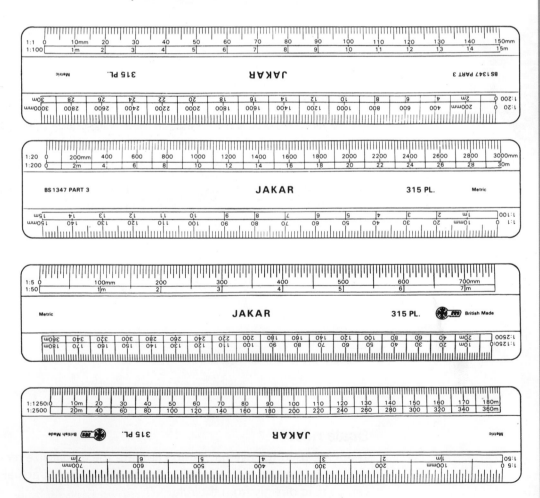

Figure 3.5

Use a scale rule to measure the following lines and then complete the table.

Figure 3.6

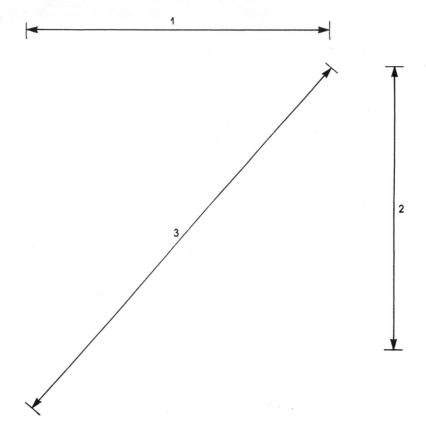

Line	Scale ratio	Length represented
1	1:1 1:100 1:50	
2	1:5 1:100 1:2500	
3	1:20 1:200 1:1250	

The outline ground floor plan of a house is drawn to a scale of 1:50.

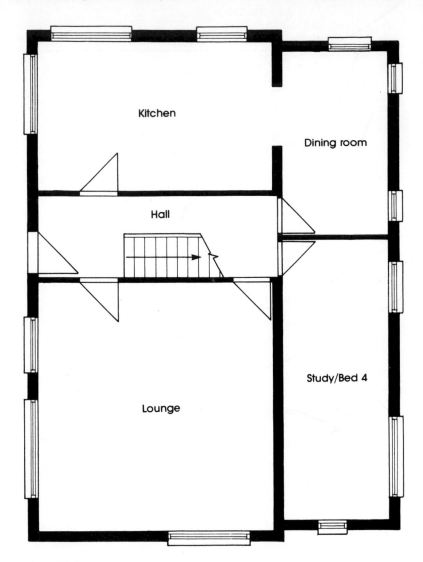

Figure 3.7

Use your scale rule to fill in the table.

Room	Length	Width	Length of skirting
Lounge			
Dining room			
Kitchen			
Study/Bed 4			
Hall			

Dimensions shown on drawings

Dimensions shown on scale drawings normally include the symbol for the measurement units used. A lower case letter 'm' is used for metres and lower case letters 'mm' for millimetres. Units should not be mixed on drawings, 2½ can be shown either as '2.5 m', '2.50 m', '2.500 m' or '2500 mm'. Units are not always included with the dimension, in these cases whole numbers are taken as millimetres (e.g. 1500) and decimal numbers to one, two or three places are taken as metres (1.5 or 1.50 or 1.500). Note that 1500 mm = 1.5 metre (or 1.50 or 1.500 metre).

On some drawings, to avoid confusing or missing the position of the decimal point, an oblique stroke is used to separate metres from millimetres e.g. 1/500 for 1 m and 500 mm (1.500 metre).

Where the dimension is less than a metre, a nought is added before the stroke, e.g. 0/750 for 750 mm.

Sequence of dimensioning

The recommended sequence for expressing dimensions is illustrated in Figure 3.8. Length is normally expressed first, followed by width and then thickness:

$$L \times W \times T = 2100 \times 225 \times 50 \text{ mm}$$

Where only the sectional size is quoted, width is normally stated before thickness, e.g. 225 x 50 mm. Beware that the sectional size on drawings may be stated with the size seen in plan first, e.g. a 100 x 50 mm section may be shown as 50 x 100 mm when used as a joist or 100 x 50 mm for a wall plate.

Lines used on drawings

A variety of different line types and thickness or 'weight' is used on drawings for specific purposes; these are illustrated in Figure 3.9.

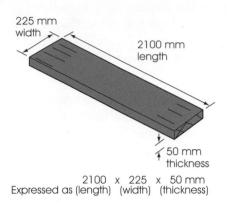

225 mm width
2100 mm length
50 mm thickness

2100 x 225 x 50 mm
Expressed as (length) (width) (thickness)

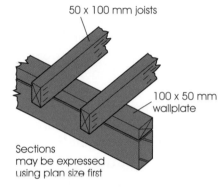

50 x 100 mm joists

100 x 50 mm wallplate

Sections may be expressed using plan size first

Figure 3.8

Sequencing dimensions

Line type		Used for
————————	Thick	Main outlines
————————	Medium	General details and outlines
————————	Thin	Construction and dimension lines
——⋀——	Break-line	Breaks in the continuity of a drawing
—·—·—·—	Thick chain	Pipe lines, drains and services
—·—·—·—	Thin chain	Centre lines
▲————▲	Section line	Showing the position of a cut (the pointers indicate the direction of view)
- - - - - - -	Broken line	Showing details which are not visible
├——┤	Dimension line	Showing the distance between two points

Figure 3.9

Line types and weights

Communications

Chapter 3

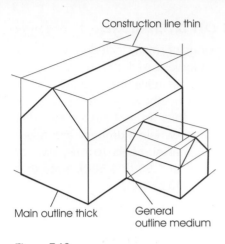

Construction line thin

Main outline thick

General outline medium

Figure 3.10

Use of line weights

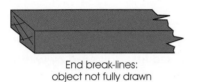

End break-lines: object not fully drawn

750 mm

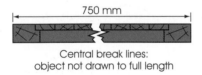

Central break lines: object not drawn to full length

Figure 3.11

Use of break-lines

◆ **Solid lines** – used for the actual parts of an object that can be seen. Thick or dark lines for the outside border; medium or lighter lines for internal borders and general detail; thin or light lines for construction and dimensioning. Figure 3.10 shows a simple building outline that illustrates the use of these different line weights.

◆ **Break-lines in a zig-zag pattern** – used to show a break in the continuity of a drawing. End break-lines are used to indicate that an object has not been fully drawn. Central break-lines are used to indicate that the object has not been drawn to its full length. See Figure 3.11.

◆ **Chain lines in a long/short or dot/dash pattern** – used for centre lines and services as illustrated in Figure 3.12.

◆ **Section lines** – shown on a drawing to indicate an imaginary cutting plane, at a particular point through an object. Pointers or arrows on the line indicate the direction of view that will be seen on a separate section drawing. Where more than one section is shown these will be labelled up as A—A, B—B etc., according to the number of sections. The section drawings themselves may be included on the same drawing sheet or be cross-referenced to another drawing (see Figure 3.13).

◆ **Broken lines** – indicate hidden details that cannot be seen on the object as drawn, or for work that is to be removed, as illustrated in Figure 3.14.

◆ **Dimension lines** – used to indicate the distance between points on a drawing. These are lightly drawn solid lines with arrowheads terminating against short cross-lines. Open (birdsfeet) arrowheads are often used for the basic/modular 'unfinished' distances in a drawing. Closed (solid) arrowheads are the preferred method for general use and 'finished' work sizes (see Figure 3.15).

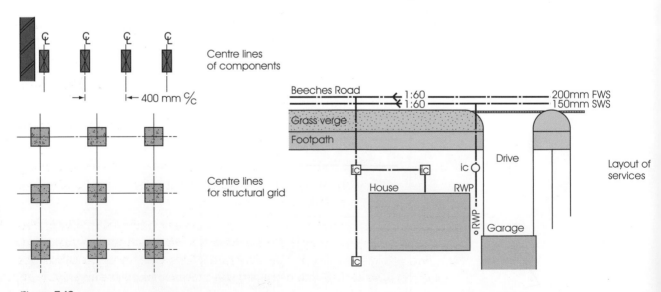

Centre lines of components

400 mm %c

Centre lines for structural grid

Beeches Road

1:60
1:60

200mm FWS
150mm SWS

Grass verge

Footpath

Drive

ic

ic

ic

RWP

House

RWP

Garage

ic

Layout of services

Figure 3.12

Chain lines

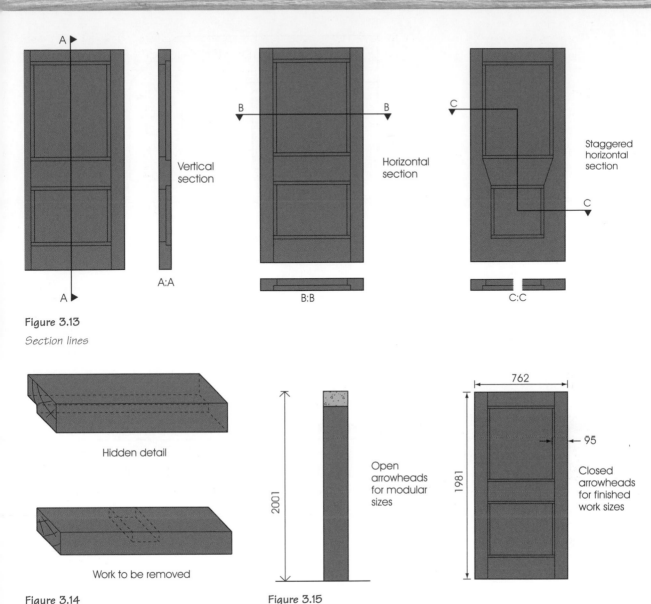

Figure 3.13
Section lines

Vertical section

A:A

Horizontal section

B:B

Staggered horizontal section

C:C

Hidden detail

Work to be removed

Figure 3.14
Broken lines

Open arrowheads for modular sizes

2001

Figure 3.15
Dimension lines

762

95

1981

Closed arrowheads for finished work sizes

Separate or running dimensions

The actual figured dimensions may be shown as either separate or running dimensions, as shown in Figure 3.16:

◆ *Separate dimensions* – normally written above and centrally along the line; figures on vertical lines should be written parallel to the line, so that they are read from the right.

◆ *Running dimensions* – shown cumulatively commencing with zero from a fixed point or datum. These are often written on the cross-line at right angles to the dimension line. Care must be taken not to confuse separate and running dimensions when reading them off a drawing.

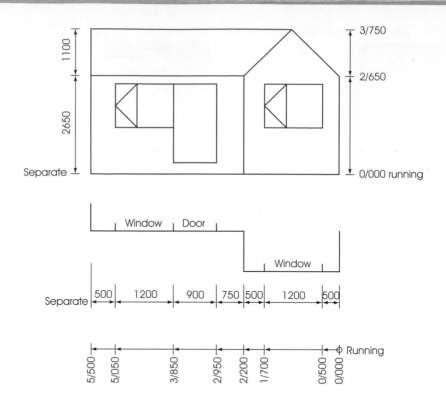

Figure 3.16

Separate and running dimensions

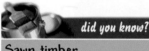

did you know?

Sawn timber
may be termed as unwrot meaning unplaned or unworked

Planed timber
may be termed as wrot meaning planed or worked.

Symbols and abbreviations used on drawings

Symbols are graphical illustrations used to represent different materials and components in building drawings. (See Figure 3.17 and 3.18)

Abbreviations are a short way of writing a word or group of words. They allow maximum information to be included in a concise way. Here are some of the abbreviations commonly used in the construction industry.

Aggregate	agg	Column	col	Joist	jst
Air brick	AB	Concrete	conc	Mild steel	MS
Aluminium	al	Copper	Copp cu	Pitch fibre	PF
Asbestos	abs	Cupboard	cpd	Plasterboard	pbd
Asbestos cement	absct	Damp proof course	DPC	Polyvinyl acetate	PVA
Asphalt	asph	Damp proof membrane	DPM	Polyvinylchloride	PVC
Bitumen	bit	Discharge pipe	DP	Rainwater head	RWH
Boarding	bdg	Drawing	dwg	Rainwater pipe	RWP
Brickwork	bwk	Expanding metal lathing	EML	Reinforced concrete	RC
BS* Beam	BSB	Foundation	fdn	Rodding eye	RE
BS Universal beam	BSUB	Fresh air inlet	FAI	Foul sewers	FS
BS Channel	BSC	Glazed pipe	GP	Sewers surface water	SWS
BS equal angle	BSEA	Granolithic	grano	Softwood	swd
BS unequal angle	BSUA	Hardcore	hc	Tongue and groove	T & G
BS tee	BST	Hardboard	hdbd	Unglazed pipe	UGP
Building	bldg	Hardwood	hwd	Vent pipe	VP
Cast iron	CI	Inspection chamber	IC	Wrought iron	WI
Cement	ct	Insulation	insul		
Cleaning eye	CE	Invert	inv		

*BS = British Standard

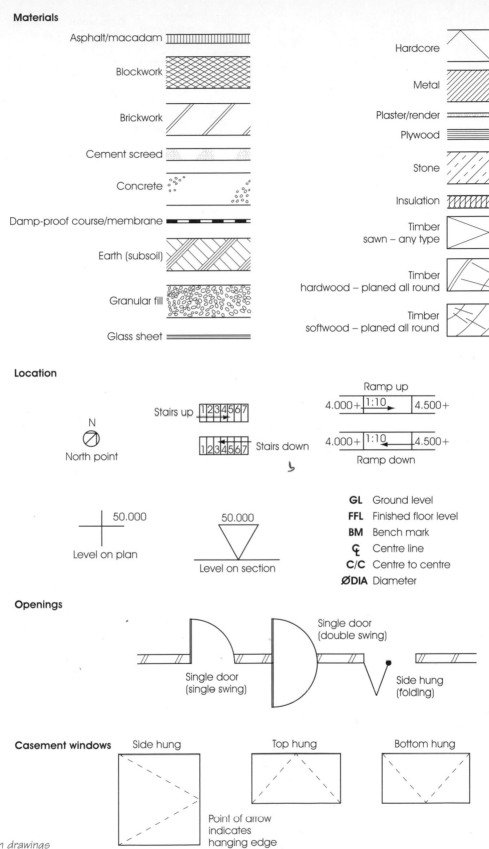

Figure 3.17

Symbols used on drawings

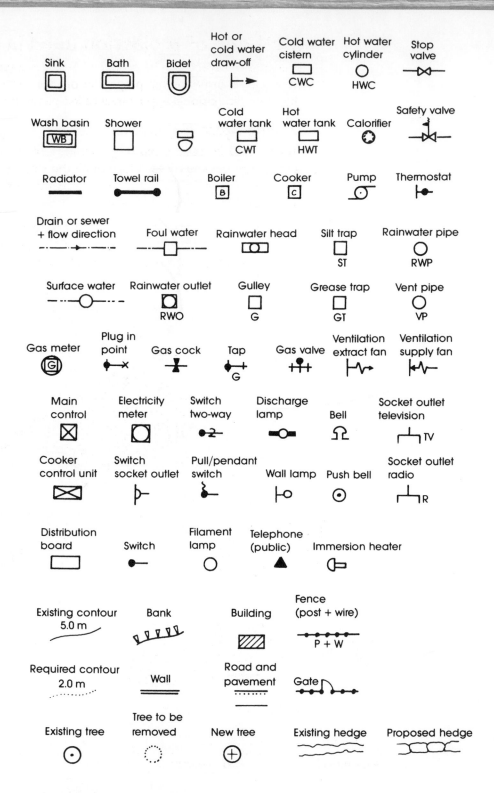

Figure 3.18

Symbols used on drawings

Methods of projection used in drawings

Drawings of objects can be produced as either a series of plan or elevation views called orthographic projection or in a form that closely resembles the three-dimensional appearance called pictorial projection.

Orthographic projection

This method is used for working drawings, plans, elevations and sections. A separate drawing of each face of all the views of an object is produced in a systematic manner on the same drawing sheet (see Figure 3.19). Each view is at right angles to the face:

◆ *Plan view* – details the surface of an object when looking down on it vertically. Floor plans in building drawings are normally drawn as a section taken just above windowsill height.

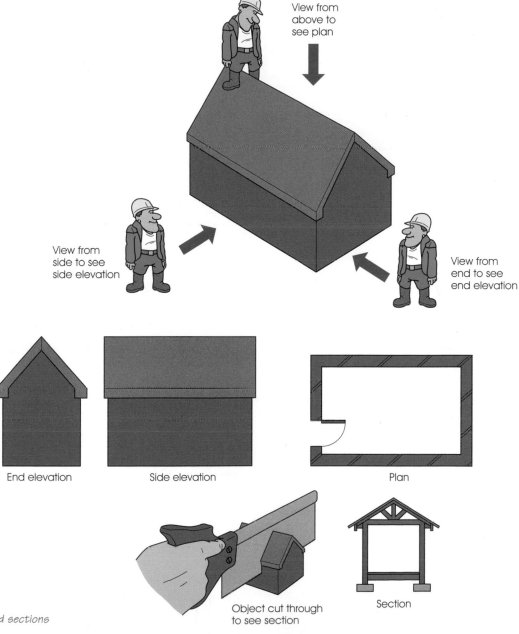

View from above to see plan

View from side to see side elevation

View from end to see end elevation

End elevation   Side elevation   Plan

Object cut through to see section

Section

Figure 3.19

Plans, elevations and sections

Chapter 3 Communications

did you know?

1st angle projection is normally used for buildings and 3rd angle for engineering drawings.

◆ *Elevation view* – of an object details the surface from side, front or rear.

◆ *Section view* – details the cut surface produced when an object is imagined cut through with a saw.

First and third angle projection

The actual position on the drawing sheet of the plan, elevations and section will vary depending on the method of projection used, either first or third angle. The viewing position in relationship to the object is illustrated in Figure 3.20.

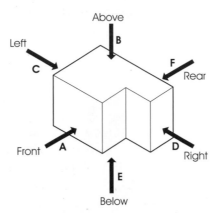

Figure 3.20

Views for orthographic projection

◆ *First angle projection* – (Figure 3.21) is generally used for building drawings where in relation to the front view the other views are arranged as follows: the view from above is drawn below; the view from below is drawn above; the view from the left is drawn to the right; the view from the right is drawn to the left; the view from the rear is drawn to the extreme right. A sectional view may be drawn to the left or the right where space permits.

◆ *Third angle projection* – (Figure 3.22) is a form of orthographic projection used for engineering drawings. It is also termed American projection. In relation to the front elevation the other views are arranged as follows: the view from above is drawn above; the view from below is drawn below; the view from the left is drawn to the left; the view from the right is drawn to the right; the view from the rear is drawn to the extreme right. A sectional view may be drawn to the left or the right where space permits.

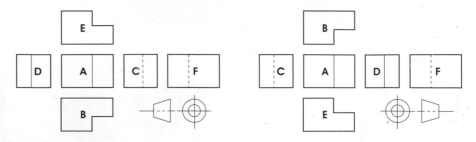

Figure 3.21

First angle projection

Figure 3.22

Third angle projection

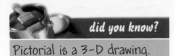

Pictorial projection

This is a method of drawing objects in a three-dimensional form. Often used for design and marketing purposes as the finished appearance of the object can be more readily appreciated by the general public. Differing views are achieved by varying the angles of the base lines and the scale of the side projections.

▸ **Isometric** – (Figure 3.23) the most widely used form of pictorial drawing where all verticall ines are drawn vertical, and all horizontal lines are drawn at an angle of 30 degrees to the horizontal. The length, width and height are all drawn to the same scale.

▸ **Planometric** – (Figure 3.24) a form of pictorial drawing where all vertical lines are drawn vertical, horizontal lines on the front elevation are drawn at 30 degrees and those on the side elevation at 60 degrees to the horizontal, giving a true plan shape. The length, width and height are all drawn to the same scale.

▸ **Axonometric** – (Figure 3.25) a form of pictorial drawing similar to planometric, except that the horizontal lines in both the elevations are drawn at 45 degrees to the horizontal, again giving a true plan shape. The length, width and height are all drawn to the same scale.

▸ **Oblique** – (Figure 3.26) a form of pictorial projection, which uses a true front elevation; the side elevation can be either **cabinet** or **cavalier**. All vertical lines are drawn vertical and all horizontal lines in the front elevation are drawn horizontal, while all horizontal lines in the side elevations are drawn at 45 degrees to the horizontal. In cavalier these 45-degree lines are drawn to their full or scale length, while in cabinet they are drawn to half their full or scale length to give a less distorted view.

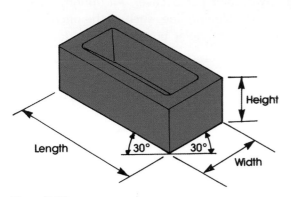

Figure 3.23

Isometric projection

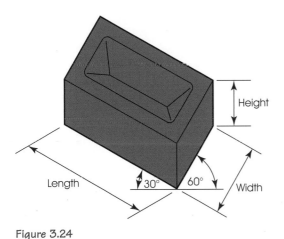

Figure 3.24

Planometric projection

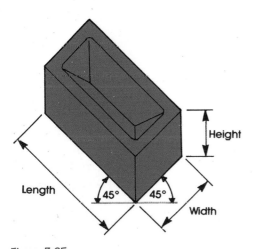

Figure 3.25

Axonometric projection

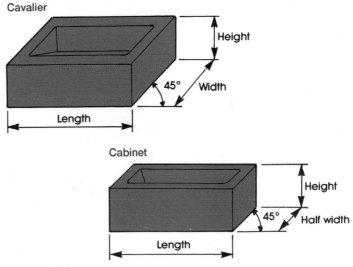

Figure 3.26

Oblique projection

did you know?

You should always check the date of a drawing to ensure you are using the latest edition. Revisions may be indicated by a letter following the drawing number, e.g. 14/0351 rev B etc.

◆ *Perspective* – a form of pictorial projection where the horizontal lines disappear to one or more points on the imaginary horizon. These disappearing points are termed as VPs or 'vanishing points'. **Parallel perspective** (Figure 3.27) has a true front elevation with the sides disappearing to one vanishing point (one-point perspective). **Angular perspective** (Figure 3.28) is drawn with the elevations disappearing to two vanishing points (two-point perspective). All upright lines in both forms are drawn vertically. Sloping lines such as the pitch of the roof are best first drawn in a rectangular box, with the ridge being positioned mid-way across the box.

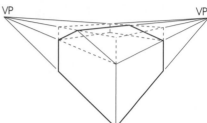

Figure 3.28

Angular (two-point) perspective projection

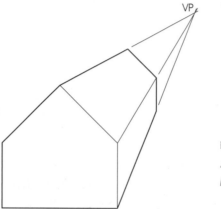

Figure 3.27

Parallel (one point) perspective projection

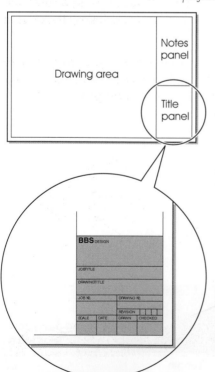

Figure 3.29

Layout of working drawing sheet

Layout and production of drawings

Drawings can be produced either by hand drafting using a drawing board and instruments or on a computer using a CAD (computer-aided design) programme. In addition to the actual labelled drawings they will also include figured dimensions, printed notes to explain exactly what is required, and a title panel, which identifies and provides information about the drawing (see Figure 3.29).

Paper sizes

Drawings are normally produced on a range of international paper sizes as illustrated in Figure 3.30 . 'A0' is the base size consisting of a rectangle having an area of 1 square metre and sides, which are in the proportion of $1:\sqrt{2}$. All the A series are of this proportion, enabling them to be doubled or halved in area and remain in the same proportion, which is useful for photographic reproduction. A1 is half A0; A2 is half A1; A3 is half A2; A4 is half A3. Where sizes larger than A0 are required the A is proceeded by a number 2A being twice the size of A0 and 4A being four times the size of A0, etc.

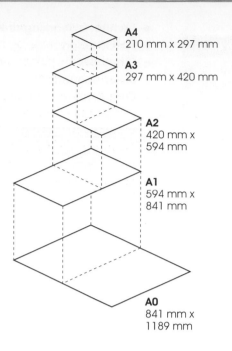

Figure 3.30

International paper sizes

A4
210 mm x 297 mm

A3
297 mm x 420 mm

A2
420 mm x 594 mm

A1
594 mm x 841 mm

A0
841 mm x 1189 mm

Sketches

These are rough outlines or initial drafts of an idea before full working drawings are made. Alternatively sketches may be prepared to convey thoughts and ideas. It is often much easier to produce a sketch of your intentions rather than to describe in words or produce a long list of instructions. Sketches can be produced either freehand, that is without the use of any equipment, or be more accurately produced using a ruler and set square to give basic guidelines. Methods of projects follow those used for drawing, e.g. orthographic or pictorial.

Pictorial sketching is made easy if you imagine the object you wish to sketch with a three-dimensional box around it. Draw the box first, lightly, with a 2H pencil, and then draw in the object using a HB pencil.

Use the box method to sketch a hand tool and a component or element associated with your occupation.

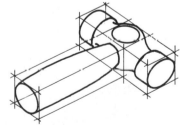

Construction activity documents

These are also known as **contract documents** and will vary depending on the nature of the work, but will normally consist of the following (see Figure 3.31):

Communications **Chapter 3**

did you know?

All contract documents must be studied to show the full extent of the work.

- ◆ Architects' working drawings
- ◆ Specification
- ◆ Schedules
- ◆ Bill of quantities
- ◆ Conditions of contract.

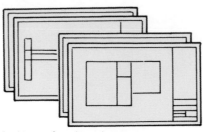

Architects' working drawings

Specification

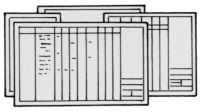

Schedules

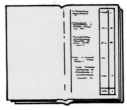

Bill of quantities

Conditions of contract

Figure 3.31

Contract documents

Figure 3.32

Block plan

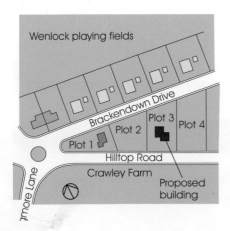

Architects' drawings

A range of scale working drawings, showing plans, elevations, general arrangements, layouts and details of a proposed construction; the main types are as follows.

Location drawings

Block plans (Figure 3.32) identify the proposed site in relation to the surrounding area. The scales most commonly used are 1:2500 and 1:1250.

Site plans (Figure 3.33) show the position of a proposed building and the general layout of the road services and drainage etc. on the site. The scales most commonly used are 1:500 and 1:200.

General location plans (Figure 3.34) show the positions occupied by the various areas within a building and identify the locations of principal elements and components. The scales most commonly used are 1:200, 1:100 and 1:50.

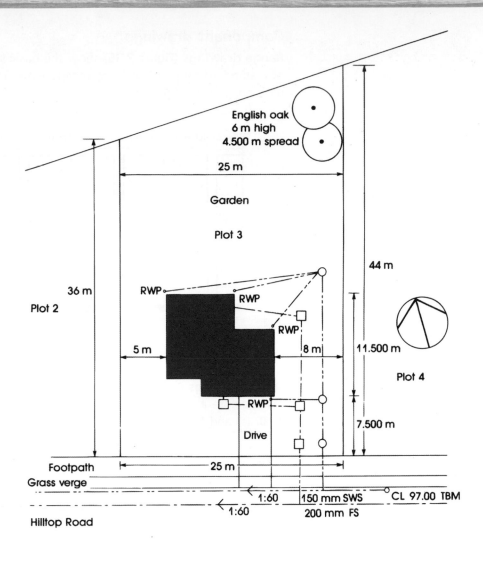

Figure 3.33

Site plan

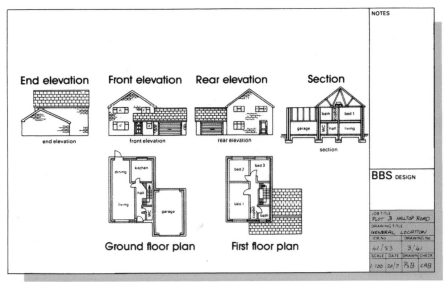

Figure 3.34

General location plan

Component drawings

Range drawings (Figure 3.35) show the basic sizes in a reference system of a standard range of building components. The scales most commonly used are 1:100, 1:50 and 1:20.

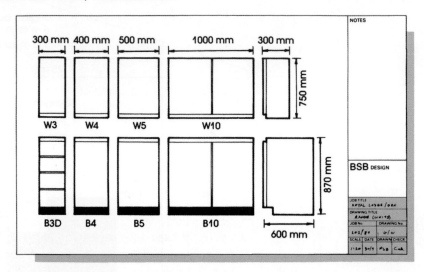

Figure 3.35

Range drawing

Detail drawings (Figure 3.36) show all the information that is required to manufacture a particular component. The scales most commonly used are 1:10, 1:5 and 1:1.

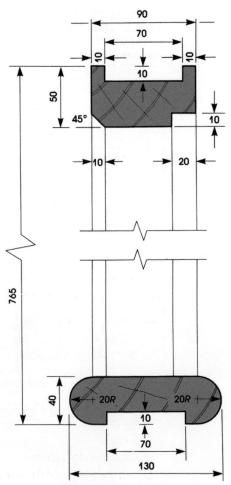

Figure 3.36

Detail drawing

Assembly drawings

Assembly details (Figure 3.37) show the junctions between the various elements and components of a building. The scales most commonly used are 1:20, 1:10 and 1:5.

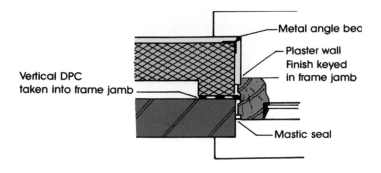

Vertical DPC
taken into frame jamb

Metal angle bea

Plaster wall
Finish keyed
in frame jamb

Mastic seal

Figure 3.37

Assembly detail drawing

Specification

Except in the case of very small building works the drawings cannot contain all the information required by the builders, particularly concerning the required standards of materials and workmanship. For this purpose the architect will prepare a document, called the specification, to supplement the working drawings. The specification is a precise description of all the essential information and job requirements that will affect the price of the work but cannot be shown on the drawings. Typical items included in specifications are:

- Site description
- Restrictions (limited access and working hours etc.)
- Availability of services (water, electricity, gas, telephone)
- Description of materials, quality, size, tolerance and finish
- Description of workmanship, quality, fixing and jointing
- Other requirements: site clearance; making good on completion; nominated suppliers and subcontractors; who approves the work, etc.

Specifications may refer to the relevant British Standards for the quality of work and materials. Various clauses of a typical specification are shown in Figure 3.38).

Schedules

These are used to record repetitive design information about a range of similar components. The main areas where schedules are used include:

- Doors, frames, linings
- Windows
- Ironmongery
- Joinery fitments
- Sanitary ware, drainage
- Heating units, radiators
- Finishes, floor, wall, ceiling
- Lintels
- Steel reinforcement
- List (register) of drawings for a project.

Figure 3.38

Extracts from a specification

BBS DESIGN

Specification of the works to be carried out and the materials to be used in the erection and completion of a new house and garage on plot 3, Hilltop Road, Brackendowns, Bedfordshire, for Mr W. Whiteman, to the satisfaction of the architect.

1.00   General conditions

1.01
1.02

1.03
1.04

2.00
2.01
2.02

2.03
2.04
2.05

2.06
2.07
2.08
2.09

10.00   Woodwork

10.01   Timber for carcassing work to be machine strength graded class C16

10.02   Timber for joinery shall be a species approved by the architect and specified as J2

10.03   Moisture content of all timber at time of fixing to be appropriate to the situation and conditions in which it is used. To this effect all timber and components will be protected from the weather prior to their use.

10.04

10.05

10.06

10.18   Construct the first floor using 50 mm × 195 mm sawn softwood joists at 400 mm centres supported on mild steel hangers.

Provide 75 mm × 195 mm trimmer and trimming around stairwell, securely tusk-tenoned together.

Provide and fix to joists 38 mm × 38 mm sawn softwood herring-bone strutting at 1.8 m maximum intervals.

Provide and fix galvanized restraint straps at 2 m maximum intervals to act as positive ties between the joists and walls.

10.19   Provide and secret fix around the trimmed stairwell opening a 25 mm Brazilian mahogany apron lining, tongued to a matching 25 mm × 100 mm nosing.

10.20   Provide and lay to the whole of the first floor 19 mm × 100 mm prepared softwood tongued and grooved floor boarding, each board well cramped up and surface nailed with two 50 mm flooring brads to each joist. The nail heads to be well punched down.

Figure 3.39

BSI Kite Mark Symbol

Chapter 3   Communications

A typical window schedule, which includes a range drawing, along with the related floor plans is illustrated in Figure 3.40. The actual details for each window have been indicated with a tick.

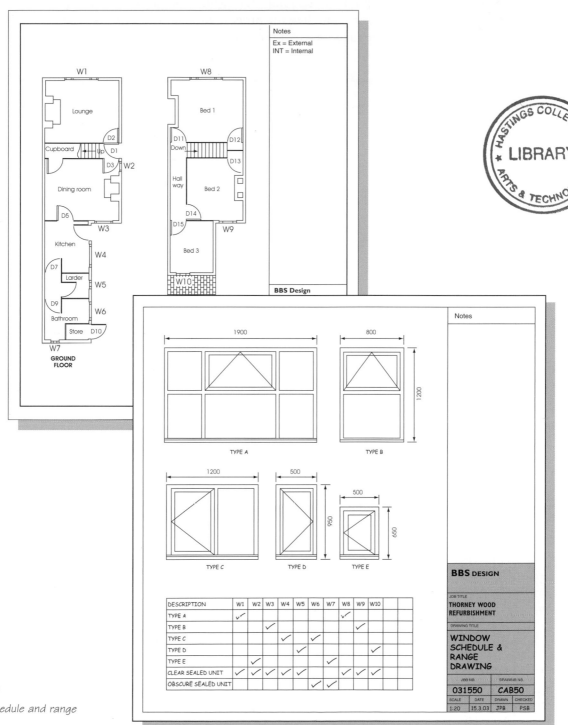

DESCRIPTION	W1	W2	W3	W4	W5	W6	W7	W8	W9	W10		
TYPE A	✓							✓				
TYPE B			✓						✓			
TYPE C				✓		✓						
TYPE D					✓					✓		
TYPE E		✓					✓					
CLEAR SEALED UNIT	✓	✓	✓	✓	✓			✓	✓	✓		
OBSCURE SEALED UNIT						✓	✓					

Notes

Ex = External
INT = Internal

BBS Design

BBS DESIGN

JOB TITLE
THORNEY WOOD REFURBISHMENT

DRAWING TITLE
WINDOW SCHEDULE & RANGE DRAWING

JOB NO.	DRAWING NO.
031550	**CAB50**

SCALE	DATE	DRAWN	CHECKED
1:20	15.3.03	JPB	PSB

Figure 3.40

Building schedule and range drawing

Communications **Chapter 3**

activity

Using the schedule for sanitary appliances, estate plan and house plan, complete the order/requisition form for the sanitary appliances required for plot number 6.

BBS DESIGN

JOB TITLE	Lakeside Estate
DRAWING TITLE	Schedule for sanitary appliances
JOB NO.	
DRAWING NO.	
SCALE	
DATE	
DRAWN	
CHECKED	

Notes:

Schedule for sanitary appliances (× denotes required). Columns = appliance/item; rows = plot and room.

Plot / Room	Inset sink	Waste disposal unit	Close Couple WC	Bidet	Pedestal wash basin	Wall hung corner basin	Bath	Shower tray	Anne	Sarah	James	Single drainer	Double drainer	Penthouse Red	Indian ivory	Honeysuckle	White	Chrome plated	Gold plated
Plot 12 En-suite			×	×	×			×	×							×		×	
Plot 12 Bath			×		×		×			×						×	×	×	
Plot 12 Cloaks			×			×					×					×	×	×	
Plot 12 Kitchen	×												×					×	
Plot 10 En-suite			×	×	×			×	×						×				×
Plot 10 Bath			×		×		×			×					×			×	
Plot 10 Cloaks			×			×					×					×		×	
Plot 10 Kitchen	×	×											×					×	
Plot 9 En-suite			×	×	×			×		×						×		×	
Plot 9 Bath			×		×		×			×						×		×	
Plot 9 Cloaks			×			×				×						×		×	
Plot 9 Kitchen	×											×						×	
Plot 8 En-suite			×	×	×			×	×								×		×
Plot 8 Bath			×		×		×		×								×		×
Plot 8 Cloaks			×			×			×								×	×	
Plot 8 Kitchen	×												×				×		
Plot 7 En-suite			×	×	×			×		×						×		×	
Plot 7 Bath			×		×		×			×						×		×	
Plot 7 Cloaks			×			×				×						×		×	×
Plot 7 Kitchen	×	×											×					×	×
Plot 6 En-suite			×	×	×			×	×	×					×				×
Plot 6 Bath			×		×		×		×	×					×				×
Plot 6 Cloaks			×			×				×					×			×	
Plot 6 Kitchen	×										×					×		×	×
Plot 2 En-suite			×	×	×			×		×						×		×	
Plot 2 Bath			×		×		×			×								×	
Plot 2 Cloaks			×			×					×					×		×	
Plot 2 Kitchen	×	×										×						×	

ITEM (see range); STYLE (see range); COLOUR; BRASS WORK

These are required on-site for installation on 15 March 2008 at the latest.

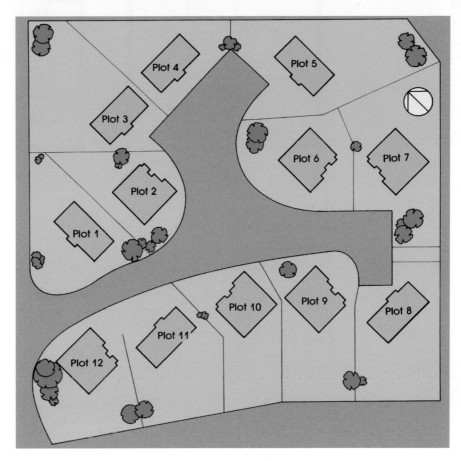

Figure 3.41

Estate plan

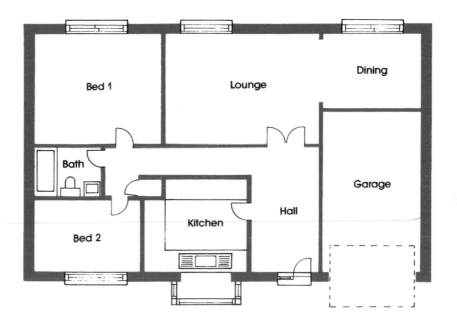

Figure 3.42

The lakeside bungalow plan

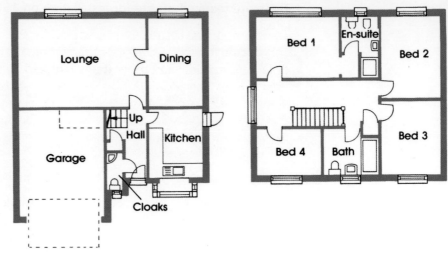

Figure 3.43

The Whiteman house plans

BBS CONSTRUCTION
ORDER/REQUISITION

Registered office

No. _____

Date _____

To _____   From _____

Address   Site address
_____   _____

_____   _____

Please supply or order for delivery to the above site the following:

Description	Quantity	Rate	Date required by

Site manager/foreman _____

Note Please advise site within 24 hours of request if order cannot be fulfilled by the date required

Figure 3.44

Bill of quantities

The bill of quantities (BOQ) is prepared by the client's quantity surveyor. This document gives a complete description and measure of the quantities of labour, material and other items required to carry out the work based on drawings, specification and schedules. Its use ensures that all estimators prepare their tender on the same information. An added advantage is that as each individual item is priced in the tender the BOQ can be used for valuing the work in progress and also forms the basis for valuing any variation to the contract.

All bills of quantities will contain the following information:

◆ *Preliminaries* – these deal with the general particulars of the work, such as the names of the parties involved, details of the works, description of the site and conditions of the contract, etc.

◆ *Preambles* – these are introductory clauses to each trade covering descriptions of the material and workmanship similar to those stated in the specifications.

◆ *Measured quantities* – a description and measurement of an item of work, the measurement being given in metres run, metres square, kilograms, etc., or just enumerated as appropriate.

◆ *Provisional quantities* – where an item cannot be accurately measured, an approximate quantity to be allowed for can be stated. Adjustments will be made when the full extent of the work is known.

◆ *Prime cost sum (PC sum)* – this is an amount of money to be included in the tender for work services or materials provided by a nominated subcontractor, supplier or statutory body.

◆ *Provisional sum* – a sum of money to be included in the tender for work that has not yet been finally detailed or for a 'contingency sum' to cover the cost of any unforeseen work.

Extracts from a typical bill of quantities are shown in Figure 3.45.

Conditions of contract

Most building work is carried out under a standard form of contract such as one of the Joint Contractors Tribunal (JCT) forms of contract. The actual standard form of contract used will depend on the following:

◆ Type of client (local authority, public limited company or private individual)
◆ Size and type of work (subcontract, small or major project, package deal)
◆ Contract documents (with or without quantities or approximate quantities).

A building contract is basically a legal agreement between the parties involved in which the contractor agrees to carry out the building work and the client agrees to pay a sum of money for the work. The contract should also include the rights and obligations of all parties and details of procedures for variations, interim payments, retention, liquidated and ascertained damages and the defects liability period.

Communications Chapter 3

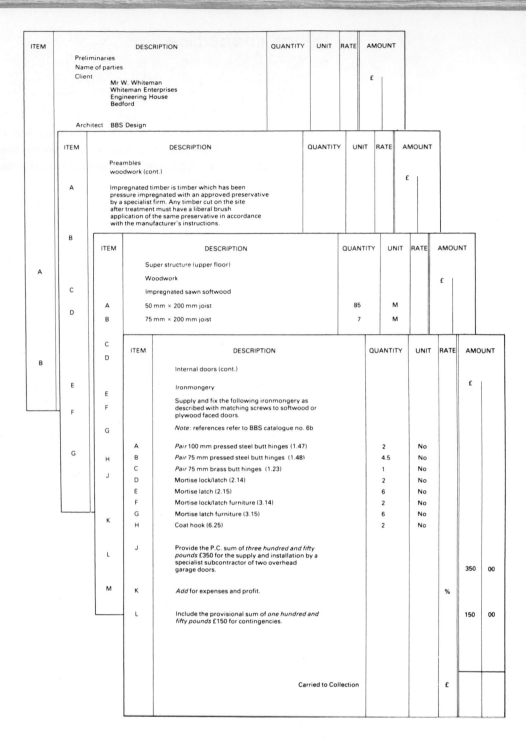

Figure 3.45

Bill of quantities

◆ **Variations** – a modification of the specification by the client or architect. The contractor must be issued with a written variation order or architect's instruction. Any cost adjustment as a result of the variation must be agreed between the quantity surveyor and the contractor.

- ◆ *Interim payment* – a monthly or periodic payment made to the contractor by the client. It is based on the quantity surveyor's interim valuation of the work done and the materials purchased by the contractor. On agreeing the interim valuation the architect will issue an interim certificate authorizing the client to make the payment.

- ◆ *Final account* – final payment on completion. The architect will issue a certificate of practical completion when the building work is finished. The quantity surveyor and the contractor will then agree the final account less the retention.

- ◆ *Practical completion* – the time at which the building work has been completed to the client's/architect's satisfaction. A certificate of practical completion will be issued and the contractor will be entitled to the remainder of monies due, less any retention.

- ◆ *Main contractor's discount* – a sum of money, which may be included in a subcontractor's quotation, often 2.5%, to cover the main contractor's administration costs associated with a subcontract.

- ◆ *Retention* – a sum of money, which is retained by the client until the end of an agreed defects liability period.

- ◆ *Liquidated and ascertained damages (LADs)* – a sum of money payable on a daily or weekly basis, agreed in advance in the contract, as being fair and reasonable payment for damages or loss of revenue to the client as a result in a delay of practical completion.

- ◆ *Defects liability period* – a period of normally six months after practical completion to allow any defects to become apparent. The contractor will be entitled to the retention after any defects have been rectified to the architect's satisfaction.

measuring up

1. State why scale drawings are used in the construction industry.

2. Explain the purpose of a range drawing.

3. Explain the purpose of a specification.

4. Describe using a typical situation the type of information that is shown on a schedule.

5. Explain why graphical symbols are used on drawings.

6. State **TWO** main details shown on a site plan.

7. What would 5 mm on a drawing to a scale of 1:20 actually represent full size?

8. Describe **THREE** methods of communication.

9. Explain what the abbreviation 'LADs' means and name the contract document that would give details of it.

Communications **Chapter 3**

10. The method of projection illustrated is:
    a) Isometric
    b) Planometric
    c) Oblique
    d) Orthographic.

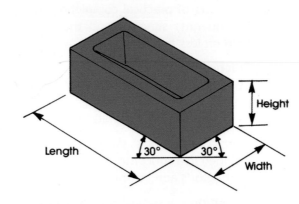

11. Produce a sketch to show the standard graphical symbol used on drawings to represent sawn timber.

12. State what a section line is used for in construction drawings.

Building control

Society exercises control over the use of land, building activities and buildings through legislation made by, or with, the authority of parliament. The three main areas of this control are:

◆ **Planning permission and development controls** restrict the type, position and use of a building or development in relation to the environment

◆ **Building regulations controls** provide functional requirements for the construction of buildings to ensure the health and safety of people in and around them, promote energy efficiency and contribute to meeting the needs of disabled people. Both of these forms of control are administered by the planning department or building control officer of the relevant local authorities, to which an application must normally be made prior to starting work

◆ **Health and safety controls** are concerned with the health and safety of all persons at their place of work and protecting other people, such as visitors and the general public, from risks occurring through work activities. The Health and Safety Executive under the Health and Safety at Work Act administers these controls.

The construction process and programme

The construction stage in the total building process can itself be subdivided into two stages: **the pre-construction stage**, which involves tendering and contract planning; and **the on-site construction stage**, which consists of the actual physical tasks and the administration processes.

The construction process can be illustrated best in the form of a flow chart to show the varied range of tasks and the order in which they are carried out. A typical flow chart of the construction process for a detached house is shown in Figure 3.46.

Tendering for a construction contract

The first of the pre-construction stages is tendering. The main way a building contractor obtains work is via the preparation and submission of tenders. There are three methods of tendering in common use: open tendering; selective tendering; negotiated contracts:

◆ *Open tendering* – architects place advertisements in newspapers and construction journals inviting contractors to tender for a particular project. Interested contractors will apply for the contract documents, and prepare and submit a tender within a specific time period. At the close of this tender period the quantity surveyor will open all the tenders and make recommendations to the architect and client as to the most suitable contractor, bearing in mind the contractor's expertise and tender price.

◆ *Selective tendering* – architects establish a list of contractors with the expertise to carry out a specific project and will ask them to submit tenders for it. The architect may make up this list, simply from experience of various contractors' expertise. Alternatively, advertisements may be placed in the newspapers and construction journals inviting contractors to apply to be included in a list of tenderers. From these applications, the architect will produce a shortlist of the most suitable contractors and ask them to tender. Again the quantity surveyor will open the returned tenders and make their recommendation to the architect and client.

◆ *Negotiated contracts* – here the architect selects and approaches suitable contractors and asks them to undertake the project. If the contractor is willing to undertake the project they will negotiate with the quantity surveyor to reach an agreed price.

Speculative building

In addition to tendering for work, many building contractors carry out work of a speculative nature. These are often termed speculative or 'spec' builders. This is where building contractors, on their own or as a part of a consortium, design and build a project for later sale, mainly private housing, but also offices, shops and industrial units. The building contractor is speculating or taking a chance that they will find a buyer (client) for the building.

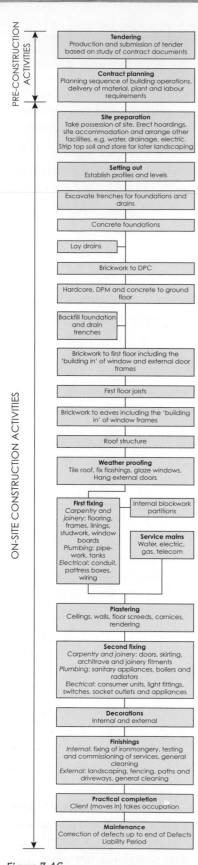

PRE-CONSTRUCTION ACTIVITIES

ON-SITE CONSTRUCTION ACTIVITIES

Tendering
Production and submission of tender based on study of contract documents

Contract planning
Planning sequence of building operations, delivery of material, plant and labour requirements

Site preparation
Take possession of site. Erect hoardings, site accommodation and arrange other facilities, e.g. water, drainage, electric. Strip top soil and store for later landscaping

Setting out
Establish profiles and levels

Excavate trenches for foundations and drains

Concrete foundations

Lay drains

Brickwork to DPC

Hardcore, DPM and concrete to ground floor

Backfill foundation and drain trenches

Brickwork to first floor including the 'building in' of window and external door frames

First floor joists

Brickwork to eaves including the 'building in' of window frames

Roof structure

Weather proofing
Tile roof, fix flashings, glaze windows. Hang external doors

First fixing
Carpentry and joinery: flooring, frames, linings, studwork, window boards
Plumbing: pipework, tanks
Electrical: conduit, pattress boxes, wiring

Internal blockwork partitions

Service mains
Water, electric, gas, telecom

Plastering
Ceilings, walls, floor screeds, cornices, rendering

Second fixing
Carpentry and joinery: doors, skirting, architrave and joinery fitments
Plumbing: sanitary appliances, boilers and radiators
Electrical: consumer units, light fittings, switches, socket outlets and appliances

Decorations
Internal and external

Finishings
Internal: fixing of ironmongery, testing and commissioning of services, general cleaning
External: landscaping, fencing, paths and driveways, general cleaning

Practical completion
Client (moves in) takes occupation

Maintenance
Correction of defects up to end of Defects Liability Period

Figure 3.46

The construction process

Contract planning and control

On obtaining a contract for a building project a contractor will prepare a programme that shows the sequence of work activities. In some cases an architect may stipulate that the contractor submits a programme of work at the time of tendering; this gives the architect a measure of the contractor's organizing ability.

The contract plan shows the interrelationship between the different tasks and also indicates when and for what duration resources such as materials, equipment and workforce are required.

Once under way the progress of the actual work can be compared with the target times contained in the plan. If the target times are realistic, the plan can be a source of motivation for the site management who will make every effort to stick to the programme and retrieve lost ground when required. There are a number of factors, some outside the management's control, which could lead to a programme modification. These factors include:

◆ Bad weather
◆ Labour shortages
◆ Labour disputes or strikes
◆ Late material deliveries
◆ Variations to contract
◆ Lack of specialist information
◆ Bad planning and bad site management, etc.

Therefore when determining the length of a contract the contractor will normally make an addition of about 10% to the target completion date to allow for such eventualities.

Bar charts

The most widely used and popular type of work plan or programme is the bar or Gantt chart. This is probably the most simple control system to use and understand. They are drawn up either manually or on a computer using a spreadsheet or specialized project planning software.

The individual tasks are listed in a vertical column on the left-hand side of the sheet and a horizontal time scale along the top. A horizontal bar or block shows the target start and end times of the individual tasks. In addition to their use as overall contract management, bar charts can be used for short-term, weekly and monthly plans (see Figure 3.47).

Various types of bar chart can be used to indicate actual progress on a contract:

◆ *Single line bar chart* – this is basically the same as the contract plan chart, but the block showing the planned activity is shown in outline and can be shaded as the actual work progresses. A sliding transparent date cursor is often laid over the chart to indicate the actual date. The progress chart shown in Figure 3.48 has the date at Thursday lunchtime. From the shading it can be seen: that both the re-wiring and re-decoration activities are behind the programme; the re-plumbing of the bathrooms is ahead of the programme.

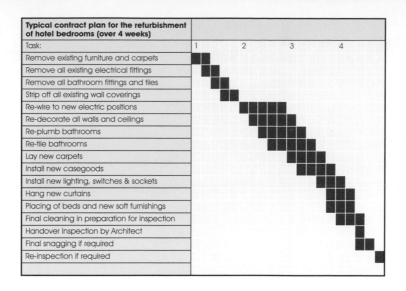

Figure 3.47

Example of a bar chart

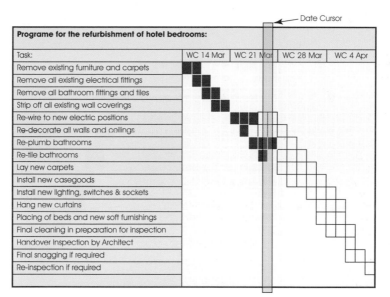

Figure 3.48

Single line bar chart

◆ **Two line bar chart** – using this type of chart the planned activity box is divided into two; one shaded to show the planned activity and the other hatched to show the actual percentage of work completed, thus making it easier to read (see Figure 3.49).

◆ **Three line bar chart** – this is basically the same as the single and two line types except that the bar is divided into three, with the third line being used to indicate the actual days worked. It can be seen from the chart (Figure 3.50) that the rewiring and re-decorating were both started a day late, thus accounting for the fact that these activities are behind the programme.

A typical bar/Gantt chart for the construction of a house is shown in Figure 3.51. This is a variation of the two line bar chart, in which the second bar is shaded in to show the work progress and the actual time taken for each task. Plant and labour requirements have also been included along the bottom of the sheet.

Communications **Chapter 3**

Figure 3.49

Two line bar chart

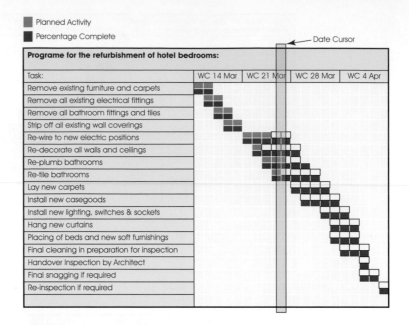

| Planned Activity |
| Percentage Complete |

Date Cursor

Programe for the refurbishment of hotel bedrooms:

Task:	WC 14 Mar	WC 21 Mar	WC 28 Mar	WC 4 Apr
Remove existing furniture and carpets				
Remove all existing electrical fittings				
Remove all bathroom fittings and tiles				
Strip off all existing wall coverings				
Re-wire to new electric positions				
Re-decorate all walls and ceilings				
Re-plumb bathrooms				
Re-tile bathrooms				
Lay new carpets				
Install new casegoods				
Install new lighting, switches & sockets				
Hang new curtains				
Placing of beds and new soft furnishings				
Final cleaning in preparation for inspection				
Handover Inspection by Architect				
Final snagging if required				
Re-inspection if required				

Figure 3.50

Three line bar chart

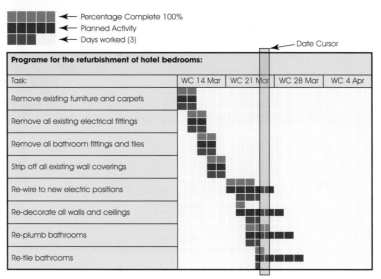

	Percentage Complete 100%
	Planned Activity
	Days worked (3)

Date Cursor

Programe for the refurbishment of hotel bedrooms:

Task:	WC 14 Mar	WC 21 Mar	WC 28 Mar	WC 4 Apr
Remove existing furniture and carpets				
Remove all existing electrical fittings				
Remove all bathroom fittings and tiles				
Strip off all existing wall coverings				
Re-wire to new electric positions				
Re-decorate all walls and ceilings				
Re-plumb bathrooms				
Re-tile bathrooms				

Annotated plans

This system of recording progress uses the actual site layout drawings, which can be hatched or coloured to a 'KEY' to show the extent of work completed to date. As with bar charts an indication of the percentage of a task completed can be indicated by the partial filling in of the appropriate section. Further activities can be included as required, using different hatching or colours (see Figure 3.52).

Numerical schedules

This method of recording progress simple shows the start and finish dates for each task and may be used in conjunction with annotated plans. From the schedule shown in Figure 3.53 it can be seen that the superstructure for example has been completed for plots 1 to 3, but whilst the superstructure for plot 4 has commenced work is still being carried out.

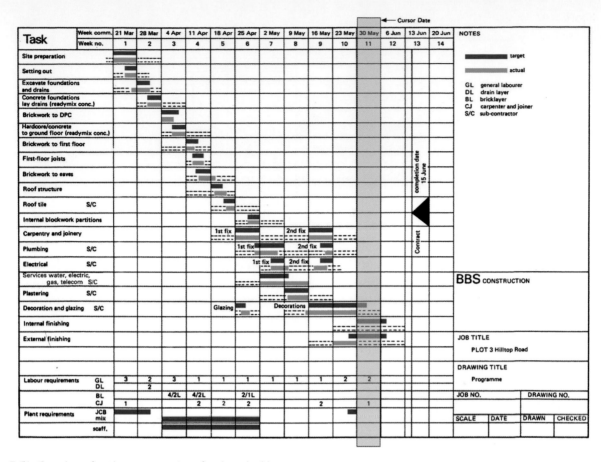

Figure 3.51 Bar chart for the construction of a detached house

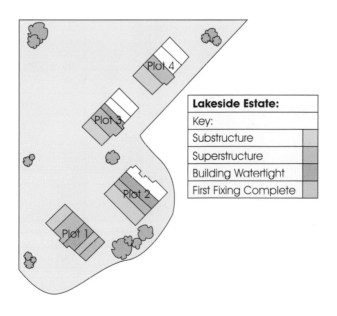

Figure 3.52

Annotated plans

Numerical Schedule: Lakeside Estate				
Task:	**Plot Number:**			
	1	2	3	4
Substructure	19/3 / 5/4	20/3 / 7/4	21/3 / 8/4	22/3 / 10/3
Superstructure	6/4 / 22/4	8/4 / 24/4	10/4 / 26/4	12/4
Building Watertight	23/4 / 26/4	25/4 / 19/3	28/4	
First Fixing Complete	30/4			

Figure 3.53

Numerical schedules

Progress graphs

This method involves inserting relevant task percentages in a computer-based spreadsheet programme and using this information to generate a graph. This results in a highly visual, simple and easily understood presentation. In addition progress is simple to update by increasing the percentages of a task completed on a periodic basis.

Progress Graph: Lakeside Estate

Task	Plot 1	Plot 2	Plot 3	Plot 4
Substructure	100	100	100	100
Superstructure	100	100	100	25
Building Watertight	100	50	35	
First Fixing Complete	35			

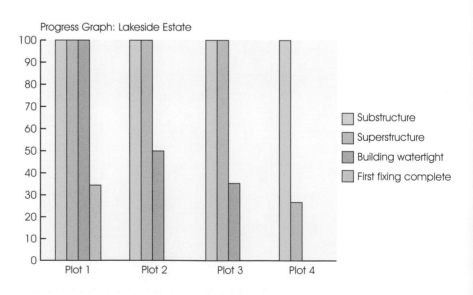

Figure 3.54

Progress graph

activity

Study the bar chart shown in Figure 3.51 and answer the following questions:

1. On what date did the work on-site commence?

2. What is the programmed completion date of the contract?

3. What week number is the contract at now?

4. To date, what activities were started before their planned date?

5. To date, what activities took longer than the target time?

6. To date, what activities took less than the target time?

7. How many carpenters were on-site during week 6 and what were they doing?

8. When was the roof tiled and who undertook the work?

9. How many weeks in total was a scaffold required?

10. What activity in week 2 was a mixer required for?

11. In what week was the glazing done, who undertook the work and was it undertaken in accordance with the programme?

measuring up

13. Name **THREE** contract documents.

14. Explain in detail the purpose or contents of one of the contract documents named in your previous answer.

15. Describe the difference between open and negotiated methods of tendering.

16. A sum of money included in a bill of quantities for work or services to be provided by a nominated subcontractor is known as the:
    a) Retention
    b) Provisional sum
    c) Prime cost sum
    d) Preambles.

17. Describe what constitutes 'first fixing' in relation to **TWO** different work roles?

Site and workshop communication

No building site or construction contract could function effectively without a certain amount of day-to-day paperwork and form filling. Methods of communication, which enable information to flow both within and between organizations include the following.

Time sheets

Each employee completes these on a weekly basis (see Figure 3.55), giving details of their hours worked and a description of the job or jobs carried out. Time sheets are used by the employer to determine wages and expenditure, gauge the accuracy of target programmes, provide information for future estimates and form the basis for claiming daywork payments. Especially on larger sites where a time clock is used, the foreman and timekeeper may complete these sheets.

Daywork sheets

Daywork sheets are not the same as time sheets. Daywork is work carried out without an estimate. This may range from emergency or repair work carried out by a jobbing builder to work that was unforeseen at the start of a major contract, for example, repairs, replacements, demolition, extra ground work, late alterations, etc. Daywork sheets should be completed by the contractor, authorized by the clerk of works or architect and finally passed on to the quantity surveyor for inclusion in the next interim payment. This payment is made from the provisional contingency sum included in the bill of quantities for any unforeseen work. Details of daywork procedures should be included in the contract conditions. A written architect's instruction is normally required before any work commences (see Figure 3.56).

Confirmation notice

Where architects issue verbal instructions for daywork or variations, written confirmation of these instructions should be sought by the contractor from the architect before any work is carried out. This does away with any misunderstanding and prevents disputes over payment at a later date. A typical confirmation notice is shown in Figure 3.57

Daily report/site diary

This is used to convey information back to head office and also to provide a source for future reference, especially should a problem or dispute arise later in the contract regarding verbal instructions, telephone promises, site visitors, delays or stoppages due to late deliveries, late starts by subcontractors or bad weather conditions. Like all reports its purpose is to disclose or record facts; it should therefore be brief and to the point. Many contractors use a duplicate book for the combined daily report and site diary. After filling it in, the top copy is sent to head office, the carbon copy being retained on-site. Some firms use two separate documents to fulfil the same function (see Figure 3.58).

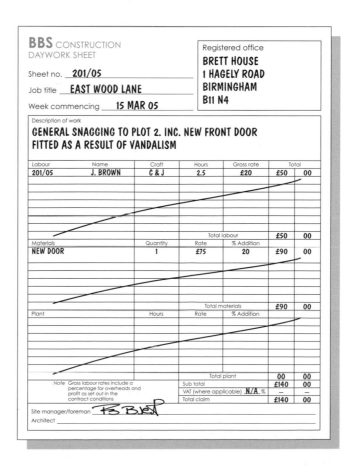

Figure 3.55
Time sheet

Figure 3.56
Daywork sheet

Communications **Chapter 3**

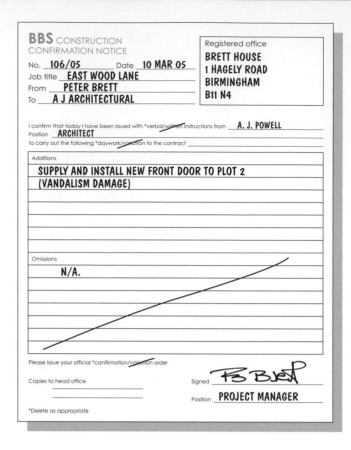

Figure 3.57

Confirmation notice

Figure 3.58

Daily report/site diary

Orders/requisitions

The majority of building materials are obtained through the firm's buyer, who at the estimating stage would have sought quotes from the various suppliers or manufacturers in order to compare prices, qualities and discounts. It is the buyer's responsibility to order and arrange phased deliveries of the required materials to coincide with the contract programme. Each job would be issued with a duplicate order/requisition for obtaining sundry items from the firm's central stores or, in the case of a smaller builder, direct from the supplier. Items of plant would be requisitioned from the plant manager or plant hirers using a similar order/ requisition (see Figure 3.59).

Delivery notes

When materials and plant are delivered to the site, the foreman is required to sign the driver's delivery note (see Figure 3.60). A careful check should be made to ensure all the materials are there and undamaged. Any missing or damaged goods must be clearly indicated on the delivery note and followed up by a letter to the supplier. Many suppliers send out an advice note prior to delivery, which states details of the materials and the expected delivery date. This enables the site management to make provision for its unloading and storage.

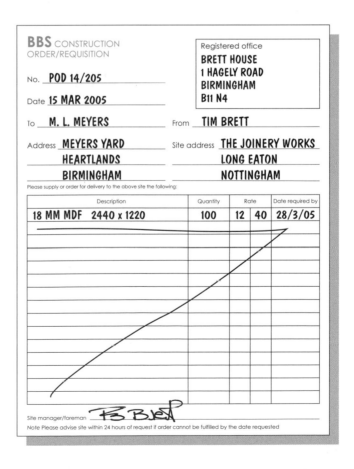

Figure 3.59

Order/requisition

Communications Chapter 3

BBS SUPPLIES
DELIVERY NOTE

No. 8914
Date 15 MARCH 2002

Registered office
BRETT HOUSE
1 HAGELY ROAD
BIRMINGHAM
B11 N4

Delivered to	Inovice to
T. JOYCEE	FELLOWS, MORTON PLC
25 DAWNCRAFT WAY	JOSHER STREET
STENSON DERBY D. 70	BIRMINGHAM B21

Please receive in good condition the undermentioned goods

SAWN, TREATED SOFTWOOD
50 OFF 25 x 50 x 3600
50 OFF 50 x 50 x 4.800
KILN SEASONED HARDWOOD
25 OFF 25 x 150 x 2400 (REBATED WINDOW SILLS)
(SHRINK-WRAPPED IN PLASTIC)

Received by *P. B. Brett*
Remarks ONLY 48 LENGTHS OF 50 x 50 RECEIVED
Note Claims for shortages and damage will not be considered unless recorded on the sheet

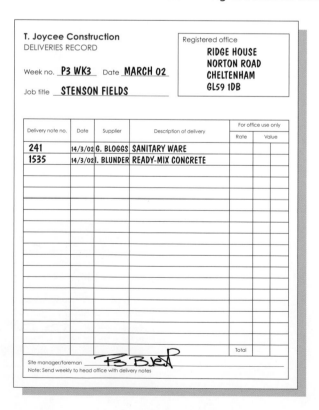

Figure 3.60
Delivery note

Delivery record

This forms a complete record of all the materials received on-site and should be filled in and sent to head office along with the delivery notes on a weekly basis (see Figure 3.61). This record is used to check deliveries before paying suppliers' invoices and also when determining the interim valuation.

T. Joycee Construction
DELIVERIES RECORD

Week no. P3 WK3 Date MARCH 02
Job title STENSON FIELDS

Registered office
RIDGE HOUSE
NORTON ROAD
CHELTENHAM
GL59 1DB

Delivery note no.	Date	Supplier	Description of delivery	For office use only	
				Rate	Value
241	14/3/02	G. BLOGGS	SANITARY WARE		
1535	14/3/02	I. BLUNDER	READY-MIX CONCRETE		
				Total	

Site manager/foreman *P. B. Brett*
Note: Send weekly to head office with delivery notes

Figure 3.61
Delivery record

Memorandum (memo)

This is a printed form on which internal communications can be carried out. It is normally a brief note about the requirements of a particular job or details of an incoming inquiry (representative/telephone call) while a person was unavailable. Internal memos written between site staff should be friendly but not frivolous, brief but factual. Carbon copies should be kept on file in case of contractual queries (see Figure 3.62).

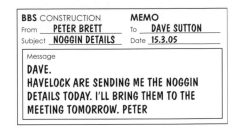

Figure 3.62

Memorandum (memo)

Tips on filling in forms

When you fill in a form for any reason, remember the following basic rules (see Figure 3.36 and 3.64):

1. Read the instructions carefully, e.g. do they ask for handwritten, BLOCK CAPITALS, or a black pen, etc.?

2. Read the questions carefully (do not squash in an answer if there is an opportunity to give that information elsewhere). Ensure your writing is legible.

3. If your name is Peter Stephen Brett, then your surname, family or last name is Brett; your forename, first name or Christian name is Peter; your other name is Stephen; your initials are PSB. Where a maiden name is asked for, this would be a married woman's surname before marriage.

4. Complete all dates, times, etc. accurately. Delete inappropriate details as asked.

5. Do not leave blanks; always write 'not applicable' or 'N/A'.

6. Do not forget to sign the form if required. This is normally your initials and surname. This is your signature.

7. Do not write where you see these:
    ◆ For official use only
    ◆ For office use only
    ◆ For store use
    ◆ For company use, etc.

8. Finally, read through the form again to ensure all sections have been completed correctly.

Figure 3.63

Form filling tips

Figure 3.64

Completed form

B CUSTOMER TO COMPLETE (BLOCK CAPITALS PLEASE)

When completing this form ensure the details show clearly on both copies.

Mr/Mrs/Ms   *P.S. BRETT*

Delivery   *70 SHALIMAR RD.*

Address   *STEPPING BROOK*

NORTHAMPTON

Full Postcode   *NN7   8WL*

Tel: Home (STD *01604*) No. *743521*

Office (STD *021*) No. *345001*

I understand that the trusses will be manufactured to the correct sizes based upon the dimensions I have provided and I accept responsibility for the dimensions.

CUSTOMER'S SIGNATURE

P B Brett

DATE *15/3/02*

Measure a window either in your home, place of work or college, etc. and use the information to complete this order form (write on a photo-copied version).

ORDER FORM

FOR MADE TO MEASURE REPLACEMENT WINDOWS

When completing this form ensure the details show clearly on all copies.

MR/MRS/MS ....
    (INITIALS)   (SURNAME)

DELIVERY ADDRESS ....

.... FULL POSTCODE ....

Please indicate where you would like the goods left, if delivered in your absence.

TELEPHONE: HOME (STD ....) No ....
                OFFICE (STD ....) No ....

I understand that the windows will be manufactured to the correct sizes based upon the dimensions I have provided and I accept responsibility for the dimensions.

CUSTOMER'S SIGNATURE .... DATE ....

STORE USE ONLY

STORE   [O] [ ] [ ]

DATE OF ORDER ....
PURCHASE ORDER No. ....
DRL No. ....
ADMIN. CHECKED BY ....
RECEIPT No. ....
TENDER TYPE: CHEQUE/CREDIT/CASH

Blue   – Order Copy
White – Store Copy
Green – Goods Inwards Copy
Yellow – Customer Copy

SEE REVERSE SIDE FOR ORDERING INSTRUCTIONS AND GUIDANCE (Enter required details and complete all sections)

SKETCH YOUR WINDOWS SHOWING DESIGNS AND DIMENSIONS HERE. N.B. ALWAYS VIEWED FROM THE OUTSIDE
(See HOW TO ORDER WINDOWS notes for opening lights min/max)
(Please note that our range of Made to Measure Windows vary in specification to our standard stock range.)

Do you require sills?   YES ☐   NO ☐
Do you intend to use these windows in conjunction with our range of standard windows?   YES ☐   NO ☐
Have you ordered made to measure windows from us before? ☐ If so, please state approx. date of order ....

Figure 3.65

Letters

These provide a permanent record of communication between organizations and individuals. They can be hand-written, but formal business letters give a better impression of the organization if they are printed. Letters should be written using simple, concise language. The tone should be polite and business-like, even if it is a letter of complaint. They must be clearly constructed with each fresh point contained in a separate paragraph for easy understanding.

When you write a letter, remember the following basic rules:

◆ Your own address should be written in full, complete with the postcode.
◆ Include the recipient details. This is the title of the person (plus name if known) and the name and address of the organization you are writing to, in full, for future reference. This should be the same as appears on the envelope.

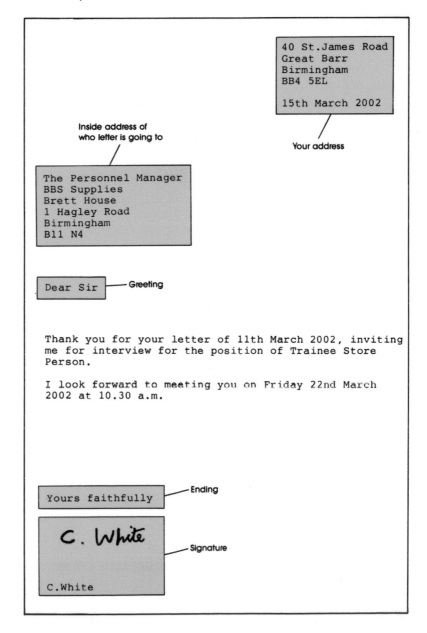

Figure 3.66

Communications Chapter 3

◆ Write the date in full, e.g. 30 January 2007.

◆ For greetings, use 'Dear Sir/Madam' if you are unsure of name and gender of the person you are writing to, otherwise 'Dear Sir' or 'Dear Madam' as applicable but use the person's name if you know it.

◆ For endings, use 'Yours faithfully' unless you have used the person's name in the greeting, in which case use 'Yours sincerely'.

◆ Sign below the ending. Your name should also be printed below the signature for clarity, with your status if appropriate.

activity

Imagine that you are an employee of T. Joycee Construction and write a letter of complaint (using a blank piece of paper or a photocopy of the headed paper below) to your materials supplier concerning the delivery you received on 15 March 2007. (See Figure 3.76 on page 141.)

T. Joycee Construction

**Ridge House
Norton Road
Cheltenham
GL59 1DB**

Figure 3.67

Facsimile transmission or FAX

Fax is a method of sending images, both text and pictures, by a telecommunications link. Most fax transmissions are via the normal telephone network. Fax machines can send and receive hand-written notes, drawings, diagrams, photographs or printed text from one fax machine to another, anywhere in the world (see Figure 3.68). Faxes are often used as a fast means of sending letters, which should normally be followed up with postal copies.

BBS Joinery Services
Brett House
1 Hagley Road
Birmingham
B11 N4

facsimile transmittal

To:	David Clarke	Fax:	01210543360
From:	Ivor Gunne	Date:	15/11/05
Re:	Standard Window Details	Pages:	2 including this one
CC:			

☑ Urgent ☐ For Review ☐ Please Comment ☐ Please reply ☐ Please recycle

David

Please find attached details of our standard range of casement windows as requested

We are also able to purpose manufacture to your requirements, should the standard range not be suitable.

Trust this information is sufficient for your purposes, however do not hesitate to contact me should you have any further queiries.

Regards

Ivor Gunne

Ivor Gunne

Production manager

Window type		Frame dimensions width x height
	V	634 x 921 634 x 1073 634 x 1226 634 x 1378
	V	921 x 921 921 x 1073 921 x 1226
	V	1221 x 1073 1221 x 1226
	CV AS CV OPP	1221 x 1073 1221 x 1226
	C AS C OPP	634 x 1073
	CD AS CD OPP	1221 x 1073
	CVC	1808 x 1073 1808 x 1226

Side hung

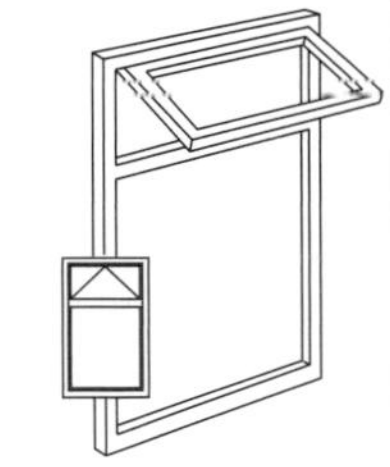

Fanlight opening

Notes:
V = Ventlight   C = Casement   OPP = Opposite hand
D = Fixed light   AS = Hand as shown

Figure 3.68

Example of a fax

E-mail or electronic mail

E-mail is the Internet's version of the postal service. Instead of faxing a message, you send a message from your computer down a telephone line. Copies of e-mails can be sent to other people who have a computer with access to the Internet and can be used as an almost instant form of memo or letter (see Figure 3.69). It is advisable to print out important e-mails to retain file copies in paper form. Electronic copies should be stored and backed up for contractual purposes.

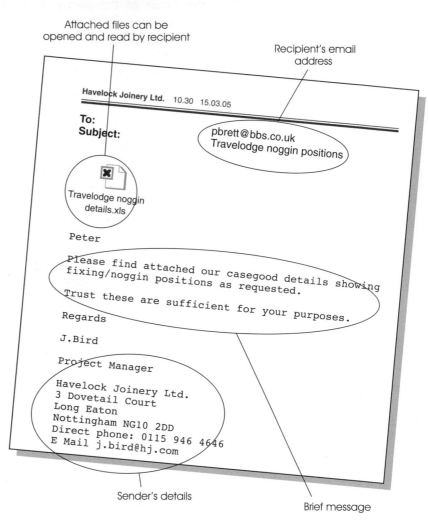

Figure 3.69

Example of an e-mail

Telephone calls

Telephones play an important communication role both within an organization and to customers and suppliers. It is useful to keep a record of incoming and outgoing telephone calls in the form of a log. Telephone manner is important. Remember that you cannot be seen and there are no facial expressions or other forms of body language to help make yourself understood. The tone, volume and pace of your voice are important. Speak clearly and loud enough to be heard without shouting; sound cheerful, with a 'smile in your voice', speaking at a speed at which the recipient can take down any message, key words or phrases that you are trying to relay.

When making telephone calls, if you initiate a call you are more likely to be in control of the conversation and when you have achieved your objective you will be in the best position to end the call without causing offence. Make notes before you begin. Have times, dates and other necessary information ready including words you find difficult to spell 'from your head'. Your part in the call may take the following form (see Figure 3.70):

◆ 'Good morning' or 'Good afternoon'
◆ 'This is (your name), of (organization) speaking'
◆ Give the name of the person you wish to speak to, if a specific individual is required
◆ State the reason for your call
◆ Keep the call brief but courteous
◆ Thank the recipient, even if the call did not produce the results required.

Figure 3.70

Making a telephone call

When receiving a telephone call, a good telephone manner is as vital as when making a call. The call may take the following form (see Figure 3.71):

◆ 'Good morning' or 'Good afternoon'
◆ 'State your organization and your name
◆ Ask 'How can I help you?'
◆ If the call is not for you and the person required is unavailable ask if you can take a message.

Figure 3.71

Receiving a telephone call

You should always make sure all messages are clear and concise to avoid any confusion.

Telephone messages

It is important that you understand what someone is saying to you on the telephone, and you may need to make notes of the conversation. When the message is not for you it is essential that you make written notes during the call, even if you are seeing the person soon and will be able to give the message verbally. Always make sure that the message contains all the necessary details, e.g. Who, Where? When? What? and How? Do not guess how to spell names and other details: ask the caller to spell them for you. Any vagueness or omission of details could lead to problems later.

Joe has to leave a message for Sid about two telephone calls. If all he writes on a scrap of paper is:

Figure 3.72

A vague message

Sid is left with the problem of finding out:

◆ How much paint each requires
◆ What colour each requires
◆ When they require it etc.

A separate message for each in the following form would be much clearer.

Telephone Message

Date 15 MARCH.. Time ...0945.................................

Message for ....SID.......................................

Message from (Name) ..JUDY..............................

(Address) ...STENSON FIELDS.........................

..............CONTRACT DERBY....................

(Telephone) .....07814 445 733.................

Message ....CAN YOU BRING TWO...............

...5 LITRE CANS OF BLUE GLOSS......

..PAINT WHEN YOU ATTEND THE.....

..SITE MEETING TOMORROW..........

.......MANY THANKS...............................

Message taken by ...JOE..........................

Figure 3.73

A clear message

At 1030 on Monday 14 March you take a telephone call for the general foreman Helen Oakes, from Vic Aston the carpentry subcontractor, based in Birmingham, telephone number 0121 354 6878.

I CAN HANG THE DOORS AT STENSON ON THURSDAY, IF YOU HAVE GOT THEM. GIVE US A BUZZ BACK TO CONFIRM. I'VE GOT THE IRONMONGERY.

Figure 3.74

On the photocopy of the form below fill in the details of this message.

Telephone Message

Date .... Time ....

Message for ....

Message from (Name) ....

(Address) ....

....

(Telephone) ....

Message ....

....

....

....

....

Message taken by ....

Figure 3.75

Two-way radio

Many large sites use a two-way radio system to enable the site foreman etc. to remain in direct communication with the site office, while roaming the site. Complete site coverage can be achieved by equipping key personnel with handsets, which are linked by a base-station located in the main site office.

Internet

The Internet is a huge collection of computers around the world, which are all linked together. These computers include those of governments, companies, trade associations, educational establishments, libraries, individuals and many more. Once you are connected you can share the information they hold, usually for free. See Introduction, 'Browsing the Internet' for further information.

activity

You have supervised the delivery of materials shown on the note below. On checking the delivery, only 48 lengths of 50 x 50 were received and several of the shrink-wrapped hardwood packages had splits in them. Using a photocopy, sign the delivery note and make any comments you think applicable.

Registered office

**Brett House
1 Hagely Road
Birmingham
B11 N4**

BBS SUPPLIES
DELIVERY NOTE

No. _8914_

Date _15 MARCH 2002_

Invoice to
_FELLOWS, MORTON PLC
JOSHER STREET
BIRMINGHAM B21_

Delivered to
_T. JOYCEE
25 DAWNCRAFT WAY
STENSON DERBY D.70_

Please receive in good condition the undermentioned goods

_SAWN TREATED SOFTWOOD
50 OFF 25 X 50 X 3600
50 OFF 50 X 50 X 4.800_

_KILN SEASONED HARDWOOD
25 OFF 25 X 150 X 2400 (REBATED WINDOW
                                            SILLS)_

(SHRINK - WRAPPED IN PLASTIC)

Received by _____
Remarks _____
Note Claims for shortages and damage will not be considered unless recorded on this sheet.

Figure 3.76

On a photocopy of the form below, complete the deliveries record for the last order.

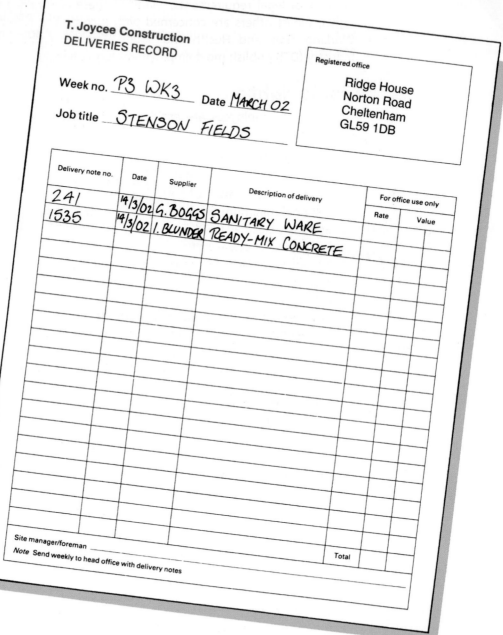

T. Joycee Construction
DELIVERIES RECORD

Week no. _P3 WK3_ Date _MARCH 02_

Job title _STENSON FIELDS_

Registered office

Ridge House
Norton Road
Cheltenham
GL59 1DB

Delivery note no.	Date	Supplier	Description of delivery	For office use only	
				Rate	Value
241	14/3/02	G. BOGGS	SANITARY WARE		
1535	14/3/02	1. BLUNDER	READY-MIX CONCRETE		

Site manager/foreman _____

Note Send weekly to head office with delivery notes   Total

Figure 3.77

Employment rights and documentation

When a company takes you into employment, it must comply with a number of legal requirements. Many of these are there to protect your rights, whilst others are concerned with Income Tax, National Insurance, Working Time and Health and Safety. The Department of Trade and Industry (DTI) publish most employment legislation.

Basic rights

Generally all employees are entitled to the following rights, which are covered by employment law:

◆ A written statement of employment
◆ A minimum of four weeks' paid annual leave
◆ Limits on the number of hours worked
◆ National minimum wage (varies with age)
◆ Statutory sick pay (SSP)
◆ Equal pay for equal work
◆ No unlawful deductions from pay
◆ Maternity pay
◆ Paternity leave
◆ Right to apply for flexible working hours
◆ Protection from discrimination on the grounds of disability, race, religion, gender or sexual preference
◆ Protection of employment upon the transfer of a business to a new owner
◆ Time off for family emergencies
◆ Time off for trade union activities
◆ Notice of termination of employment
◆ Redundancy pay
◆ A safe and healthy place of work.

Terms and conditions of employment

On appointment to a company all employees should be given a contract of employment, which give details or refers to other documents that give details of the following:

◆ Name of employer and employee
◆ Date employment began and any previous employment that will count towards continuous employment
◆ Job title or brief description
◆ Place of work
◆ Hours of work
◆ Rate of pay
◆ Overtime payments
◆ Frequency of pay
◆ Holiday entitlement including public holidays and holiday payment
◆ Pension entitlement
◆ Disciplinary and grievance procedures

did you know?

The first action to take in the case of a grievance or dispute at work is, in most circumstances, to inform your supervisor or line manager.

Communications **Chapter 3**

◆ Sickness procedures and sick pay entitlement
◆ Notice period and termination of employment
◆ Details of any collective agreements.

The Working Rule Agreement (WRA)

Most people in the construction industry are employed using the wage rates, terms and conditions of employment as laid down by the Construction Industry Joint Council (CIJC).

The Working Rule Agreement (WRA) is a collective agreement between construction industry employers and the trade unions that make up the CIJC. It is a three-year agreement that determines rates of pay, working conditions and industrial relations for the construction industry. Basic rates and main conditions are determined at national level but regional committees can negotiate variations and additions to suit a particular type of work or local practice.

Personal communications

Between yourself and work colleagues

It is necessary, in order for companies to function effectively, that they establish and maintain good working relationships within their organizational structure. This can be achieved by cooperation and communication between the various sections and individual workers: good working conditions (pay, holidays, status, security, future opportunities and a pleasant, safe working environment) are important. But so is nurturing a good team spirit, where people are motivated, appreciated and allowed to work on their own initiative under supervision for the good of the company as a whole. Your working relationships with your immediate colleagues are important to the team spirit and overall success of the company. Remember: always plan your work to ensure ease of operation, coordination and cooperation with other members of the workforce.

With customers

Ultimately it is the customer who pays your wages; they should always be treated with respect. You should be polite at all times, even with those who are difficult. Listen carefully to their wishes and pass to a higher authority in the company anything you cannot deal with to the customer's satisfaction.

Remember, when working on their property you are a guest and you should treat everything accordingly:

◆ Always treat customers' property with the utmost care
◆ Use dustsheets to protect carpets and furnishings when working indoors
◆ Clean up periodically and make a special effort when the job is complete
◆ If any problems occur contact your supervisor.

Body language

Body language is the unspoken communication that goes on in every face-to-face meeting between people. Your posture, facial expressions, eye contact and hand gestures can say more about you than your actual speech. People interpret body language all the time, although mainly on a subconscious level. You can use body language yourself when trying to make a good impression as well as reading someone else's to get an impression of whether or not they mean what the are saying.

Posture

Refers to the way you hold yourself:

◆ Turning your body towards someone shows you are attentive, more so if you lean forwards towards them at the same time
◆ Turning away or leaning back shows a lack of interest or a level of reserve
◆ Keeping your head down or hunching your shoulders indicates that you feel insignificant and do not want to draw attention to yourself
◆ Sitting with legs crossed can indicate boredom, whereas legs apart shows you are open and relaxed
◆ Folding your arms across your chest indicates defensiveness, while arms open, especially with palms uppermost, shows you are honest and accepting.

Facial expressions

The face is the most expressive part of the body:

◆ A relaxed brow shows that you a comfortable with the situation, while a tense or wrinkled brow indicates confusion, tension or fear
◆ A smile or relaxed face helps you to appear open, confident and friendly, while not smiling or a tensed face can make you appear disapproving or disinterested.

Eye contact

This gives clues to your emotions:

◆ Direct eye contact indicates that you are interested, confident and calm, however prolonged eye contact coupled with a tense face can indicate aggression
◆ Limited or no eye contact can indicate that the person is disinterested, uncomfortable or distracted
◆ Your gaze can be used to show that you are alert and interested, when walking into a room look around to give the impression that you care about where you are; keeping your eyes averted makes you appear nervous and less approachable; look up and direct your gaze around the room or the group you are with from time to time to maintain contact
◆ Making the eyes look larger, combined with an open mouth or smile shows happiness, however when combined with a tense brow or frown it can indicated that you are unhappily surprised.

Hand gestures

These can be used to emphasize other body language:

◆ Moving your hands apart with palms uppermost makes people appear more open and honest; moving your hands together can draw emphasis to what you are saying
◆ Pointing a finger can be used for emphasis, but a pointing or poking a finger at someone indicates anger
◆ Rubbing hands together shows anticipation; wringing or clenching hands shows tension; clenching the hand to make a fist shows aggression
◆ Tapping or drumming fingers indicates impatience; steepling fingers is seen as authoritative
◆ Fidgeting with hands or objects can indicate boredom, fear or difficulty in coping with a situation
◆ Avoid too many hand gestures as this can make you appear uncontrolled or nervous.

Figure 3.78

Personal hygiene

Ensure good standards of personal hygiene especially when working in occupied customers' premises. A smelly, dirty work person will make the customer assume the work will be poor. This may cause them to withdraw their offer of employment or may result in further work being given to another company. Always remember the following:

◆ Wash frequently
◆ Use deodorant if you have a perspiration problem
◆ Wear clean overalls and have them washed at least once a week
◆ Take off your muddy boots when working on customers' premises.

measuring up

18. List SIX basic rights that are covered by employment law.

19. Describe the purpose of the Working Rule Agreement.

20. What action should be taken if a materials delivery does not match up to the delivery note?

21. State the purpose of the national code in a telephone number.

22. Whom should you contact in the event of a problem occurring in a customer's home, which is outside your control?

23. Complete the memo below using a photocopy or separate piece of paper, to Christine Whiteman advising her that you will be able to attend the Safety Meeting next Friday.

BBS CONSTRUCTION   MEMO

From _____   To _____

Subject _____   Date _____

Message

Figure 3.79

24. Explain the main difference between planning permission and building regulations controls.

25. Name the document both you and your employer should sign on appointment to a company and list **THREE** things it should contain.

26. Explain the difference between a timesheet and a daywork sheet.

27. You observe the body language (illustrated) of a customer from a distance; your workmate tells you she was well happy with the work done. Do you believe them?

28. Other than a bar chart state how the progress of work on-site could be recorded.

29. List **THREE** factors that could affect the progress of work on-site.

30. State how you can maintain good working relations with:
   a) The client
   b) Your work colleagues.

Figure 3.80

Communications Chapter 3

Materials

This chapter is intended to provide the new entrant with an overview of the materials that are used in the construction industry. Its contents are assessed in the **NVQ Unit No. VR 03 Move and Handle Resources**. It is concerned with the safe handling and storage of materials that may be encountered on a day-to-day basis. You will be assessed in the context of your specific work role and environment.

In this chapter you will cover the following range of topics:

◆ On-site provision for storage of materials
◆ Storage of bulk durable building materials
◆ Storage of hazardous products
◆ Storage of fragile or perishable materials
◆ Storage of miscellaneous materials
◆ Disposal of waste.

Related information:

In addition to the contents of this chapter you will also require knowledge of the following topics to successfully complete VR 03 Move and Handle Resources:

◆ Correct selection and use of PPE
◆ Correct selection and use of fire extinguishers
◆ Regulations concerning materials, their safe handling and need for PPE
◆ Information provided by manufacturers and suppliers of materials.

These topics are covered in chapter 2, 'Health and Safety', which you should review in the context of this unit.

On-site provision for storage of materials

Site storage provision

A building site can be seen as a temporary factory, a workshop and materials store from which a contractor will construct a building. An important consideration when planning the layout of this temporary factory is the storage of materials. Their positioning should be planned in relation to where on-site they are to be used, while at the same time providing protection and security (see Figure 4.1).

Large valuable items

Frames, pipes and drainage fittings, etc. should be stored in a lockable, fully fenced compound.

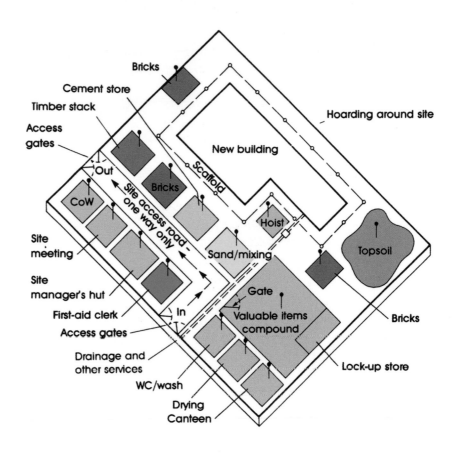

Figure 4.1
Site storage

Smaller valuable items

Carpenters' ironmongery, fixings; plumbers' copper pipe, fittings, appliances; electricians' wire, fittings; and paint, should be kept secure in one or more lockable site huts, depending on the size of the site. Like items should be stored adjacent to each other on shelving or in a bin system, as appropriate.

Each shelf or bin should be clearly marked with its contents, and each item entered on a tally card.

If materials are returned to stock they should be put back in the correct place and the tally card amended.

Heavy items should be stored at low level.

New deliveries should be put at the back of existing stock. This ensures stock is used in rotation and does not deteriorate due to an exceeded shelf-life, making it useless.

Materials Chapter 4

Issues of stores should be undertaken by a storeperson or supervisor against an authorized requisition (see Figure 4.2). Each issue should be recorded on the tally card. On some sites the tally card system (Figure 4.3) of recording the issue of materials may have been superseded by the use of a computerized system.

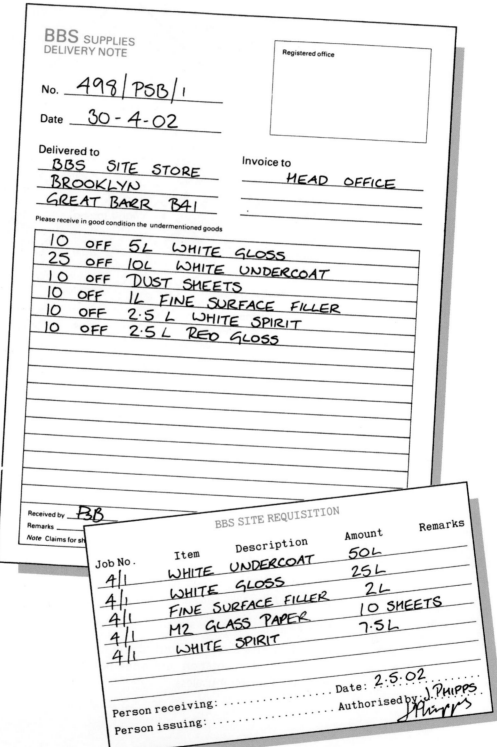

Figure 4.2

Delivery note and requisition

BBS SUPPLIERS TALLY CARD

Description of materials: WHITE GLOSS   Ref No: BSB 14/2

Size or No: 5 LITRE

Date	Order No.	Amount Inwards	Amount Outwards	Signature	Balance
15.3.02	B14	50		PBB	50
18.3.02			10	PBB	40
27.3.02			12	PBB	28
19.4.02	B75	30		PBB	58
20.4.02			16	PBB	42

Figure 4.3

Tally card

Working on photocopies, update the following tally cards to include these latest deliveries and the site requisition information. (See Figure 4.4.)

BBS SUPPLIERS TALLY CARD

Description of materials: WHITE UNDERCOAT   Ref No: BSB 15/3

Size or No: 10 LITRE

Date	Order No.	Amount Inwards	Amount Outwards	Signature	Balance
8·3·02	B11	65		BB	65
15·3·02			10	BB	55
16·3·02			1	BB	54
18·3·02			15	BB	39
27·4·02			12	BB	27

BBS SUPPLIERS TALLY CARD

Description of materials: WHITE SPIRIT   Ref No: BSB 15/50

Size or No: 2.5 LITRE

Date	Order No.	Amount Inwards	Amount Outwards	Signature	Balance
8·3·02	B12	10		BB	10
18·3·02			2	BB	8
27·3·02			3	BB	5
30·3·02	B54	10		BB	15
20·4·02			4	BB	11
28·4·02			3	BB	8

Figure 4.4

Storage requirements

Different materials have different storage requirements. These are the points to bear in mind.

Delivery dates

Phased deliveries of material should be considered in line with the planned construction programme. This will prevent unnecessarily long periods of site storage and unnecessarily large storage areas being needed.

Physical size, weight and delivery method

This will determine what plant (crane or fork-lift truck, etc.) and labour is required for off-loading and stacking.

Protection

Many materials are destroyed by extremes of temperature, absorption of moisture or exposure to sunlight, etc. Stores should be maintained as far as possible at an even temperature of about 15°C (59°F). High temperatures cause adhesives, paints, varnishes, putties and mastics, etc. to dry out and harden.

Flammable liquids such as white spirit, thinner, paraffin, petrol, some paints and varnishes, some timber preservatives and some formwork release agents must be stored in a cool, dry, lockable place. Fumes from such liquids present a fire hazard and can have an overpowering effect if inhaled. Stores of this type should always have two or more fire exits and be equipped with suitable fire extinguishers in case of fire. (See chapter 2, 'Health and Safety'.)

Water-based materials, such as emulsion paints and formwork release agents, may be ruined if allowed to freeze.

Non-durable materials such as timber, cement and plaster require weather protection to prevent moisture absorption.

Boxed or canned dry materials, such as powder adhesives, wallpaper paste, fillers, detergent powders and sugar soap quickly become useless if exposed to any form of dampness. Dampness will also rust metal containers, which may result in leakage and contamination of the contents.

Where materials are stored in a building under construction, ensure the building:

◆ has dried out after so-called 'wet trades', such as brickwork and plastering, are finished
◆ is fully glazed and preferably heated
◆ is well ventilated – this is essential to prevent the build-up of high humidity (warm, moist air).

Transit

Non-durable materials should be delivered in closed or tarpaulin-covered lorries. This will protect them from both wet weather and moisture absorption from damp or humid atmospheres.

Handling

Careless or unnecessary repeated handling will result in increased costs through damaged materials and even personal injury. Large items are best moved using two persons (see Figure 4.5).

Security

Many building materials are 'desirable' items – they will 'grow legs' and walk away if site security lapses.

Materials

Chapter 4

Figure 4.5

Handle with care!

Safety

Finally, take care of your personal hygiene. This is just as important as any physical protection measure. Certain building materials, e.g. cement, admixtures and release agents, can have an irritant effect on skin contact; they are poisonous if swallowed and can result in narcosis if their vapour or powder is inhaled. By taking proper precautions these harmful effects can be avoided. Follow manufacturers' instructions; avoid inhaling spray mists, fumes and powders; wear disposable gloves or a barrier cream; thoroughly wash hands before eating, drinking, smoking and after work. In case of accidental inhalation, swallowing or contact with skin, eyes, etc. medical advice should be sought immediately.

Figure 4.6

Take note of material labelling and manufacturers' instructions

Caution, risk   Caution, toxic   Caution, corrosive
of fire   hazard   substance

1. State the purpose of storing materials on-site.

2. Briefly describe **THREE** main points to be considered when determining on-site storage requirements.

Storage of bulk durable building materials

Bricks

These are walling unit components having a standard size, including a 10-mm mortar allowance, of 225 mm × 112.5 mm × 75 mm.

Bricks are normally made from either calcium silicate or clay. Clay bricks are usually pressed, cut or moulded and then fired in a kiln at very high temperatures. Their density, strength, colour and surface texture will depend on the variety of clay used and the firing temperature. Calcium silicate bricks are pressed into shape and steamed at high temperature. Pigments may be added during the manufacturing process to achieve a range of colours.

The three main types of bricks are as follows:

◆ **Common or Fletton bricks** are basic bricks used in the main for internal or covered (rendered or cladded) external work, although sand-faced Flettons are available for use as cheap facing bricks.
◆ **Facing bricks** are made from selected clays and are chosen for their attractive appearance rather than any other performance characteristic.
◆ **Engineering bricks** have a very high density and strength and do not absorb moisture; they are used in both highly loaded and damp conditions such as inspection chambers, basements and other sub-structure work.

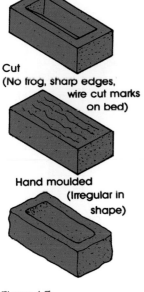

Pressed
(Regular in shape with sharp edges)

Cut
(No frog, sharp edges, wire cut marks on bed)

Hand moulded
(Irregular in shape)

Figure 4.7

Types of brick

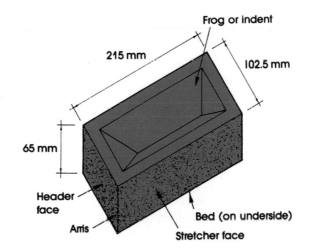

Frog or indent

215 mm

102.5 mm

65 mm

Header face

Arris

Bed (on underside)

Stretcher face

Figure 4.8

Brick terminology

Bricks may be supplied loose or banded in unit loads, shrink-wrapped in plastic and sometimes on timber pallets.

Loose bricks should be off-loaded manually, never tipped; they should be stacked on edge in rows, on level, well-drained ground. Do not stack too high: up to a maximum of 1.8 metre.

Materials

Chapter 4

Figure 4.9

Never tip bricks

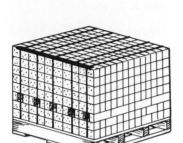

Figure 4.10

Banded load of bricks on pallet

Careless handling can chip the faces and arrises (corners), and also lead to fractures, making the bricks useless for both face and hidden work. Poor stacking creates an untidy workplace and unsafe conditions for those working or passing through the area.

Banded loads of up to 500 are off-loaded mechanically using the lorry-mounted device, a fork-lift truck or a crane. Bricks stacked on polythene or timber pallets will be protected from the absorption of sulphates and other contaminants, which could later mar the finished brickwork.

To protect bricks against rain, frost and atmospheric pollution, all stacks should be covered with a tarpaulin or polythene sheets weighted at the bottom as illustrated in Figure 4.11.

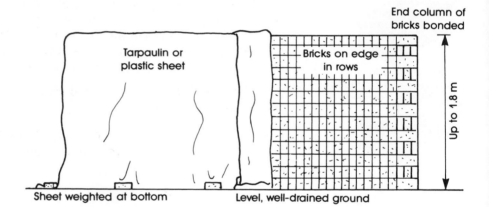

Figure 4.11

Protection of brick stack

Blocks

These are walling units that are larger than bricks, normally made either from concrete or natural stone.

Concrete blocks can be either dense or lightweight; dense blocks are often made hollow to lighten them, lightweight blocks can use either a lightweight aggregate or a fine aggregate that is aerated to form air bubbles. Concrete blocks are most often used for internal partition walls of the inner leaf of cavity walls. When used externally, they are normally either rendered (covered with a thin layer of cement mortar) or covered in cladding (tiles, slates or timber), to provide a waterproof construction. The main advantage of blocks over bricks is their increased speed of laying and also the good thermal insulation qualities of the aerated variety.

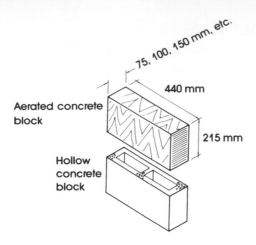

Figure 4.12

Concrete blocks

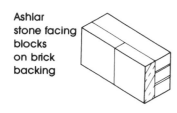

Figure 4.13

Stone blocks (ashlar)

Stone blocks are made from a naturally occurring material such as granite, sandstone, limestone, marble and slate. They are mainly used as thin-dressed stone facings known as ashlar, which are fixed to a brickwork or concrete backing.

Blocks may be either supplied loose, banded in unit loads, shrink-wrapped in plastic packs and sometimes on timber pallets.

Loose blocks should be off-loaded manually, never tipped; they should be stacked on edge in rows or columns, on level, well-drained ground. Do not stack too high: six to eight courses maximum.

Banded or palleted loads are off-loaded mechanically using the lorry-mounted device, a fork-lift truck or a crane.

To protect blocks against rain, frost and atmospheric pollution, all stacks should be covered with a tarpaulin or polythene sheets weighted at the bottom.

Stone blocks are often stored in straw, or other similar soft packing to protect arrises (corners) from impact damage.

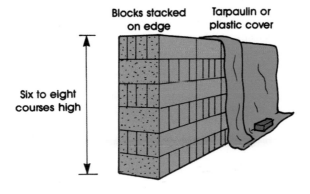

Figure 4.14

Protection of block stack

Materials

Chapter 4

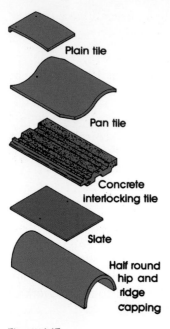

Figure 4.15

Roof tiles

Roof tiles

These are normally a terracotta clay product, a natural, metamorphic stone slate or a cast concrete product.

Roof tiles may be either supplied loose, in banded packs, in shrink-wrapped plastic packs or in unit loads on timber pallets.

Loose tiles should be off-loaded manually, never tipped; they should be stacked on edge in rows, on level, well-drained ground. Do not stack too high: four to six rows maximum; and taper the stack towards the top. End tiles in each course should be laid flat to prevent toppling. Ridge and hip cappings should be stored on end.

Banded, packed or palleted loads are off-loaded mechanically using the lorry-mounted device, a fork-lift truck or a crane.

To protect tiles against rain, frost and atmospheric pollution, all stacks should be covered with a tarpaulin or polythene sheets weighted at the bottom.

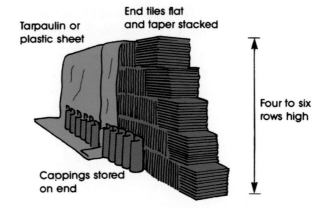

Figure 4.16

Protection of tile stack

Concrete units (paving slabs, kerbs and lintels)

These are normally pre-cast in factory conditions and transported to site.

Concrete units may be either supplied loose singly, in banded packs, in shrink-wrapped plastic packs or in unit loads on timber pallets.

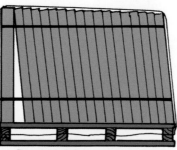

Figure 4.17

Banded paving slabs on pallet

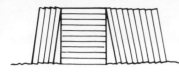

Figure 4.18

Stacking of paving slabs

Figure 4.19

Stacking of lintels

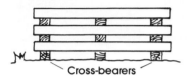

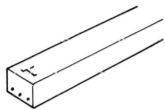

Figure 4.20

Transporting lintels

Loose concrete units should be off-loaded manually, never tipped. Items of equipment can be used to lift single units, e.g. kerb lifter.

Paving slabs should be stacked on edge in single height rows, on level, well-drained ground. Intermediate stacks of slabs laid flat can be introduced to prevent toppling.

Kerbs and lintels should be stacked flat on timber cross-bearers laid on level ground. Cross-bearers should be laid between each layer, to provide a level surface, to spread the load and to prevent the risk of distortion and chipping damage. Do not stack too high: four to six layers maximum.

Lintels are designed to contain steel reinforcement towards their bottom edge, to resist tensile forces. It is most important that they are moved in the plane of intended use, otherwise they may simply fold in two. Where the steel bars or wires cannot be seen on the end, the top edge is often marked with a 'T' or 'TOP' for identification.

Banded, packed or palleted loads are off-loaded mechanically using the lorry-mounted device, a fork-lift truck or a crane.

To protect concrete units against rain, frost and atmospheric pollution, they may be covered with a tarpaulin or polythene sheets weighted at the bottom.

Drainage pipes and fittings

These may be glazed or unglazed clay products, cast iron or UPVC (a rigid or unplasticized polyvinylchloride).

Drainage pipes and fittings may be either supplied loose singly, in banded packs, in shrink-wrapped plastic packs or in unit loads on timber pallets.

Loose pipes and fittings should be off-loaded manually, never tipped. Pipes should be stacked horizontally in rows and wedged or chocked to prevent rolling, on level, well-drained ground. Do not stack too high: 1.5 metres maximum; and taper the stack towards the top. Spigot and socket pipes should be stacked on timber cross-bearers; alternate rows should be reversed to allow sockets to project beyond spigots. Gullies and other fittings should be stacked upside down and supported so that they remain level.

Banded, packed or palleted loads are off-loaded mechanically using the lorry-mounted device, a fork-lift truck or a crane.

Materials

Chapter 4

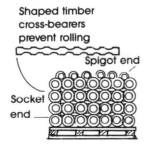

Shaped timber
cross-bearers
prevent rolling

Spigot end

Socket
end

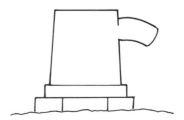

Figure 4.21 *Stacking of drainage pipes and fittings*

Aggregates

These are sands, gravel and crushed rock, which are added to cement as a filler material to produce concrete and mortar.

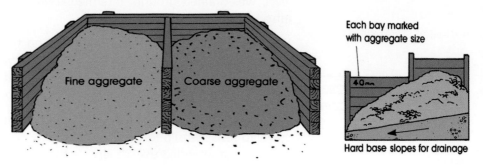

Figure 4.22 *Storage of aggregate*

Normally supplied in bulk by tipper lorries; small amounts are available bagged. Each size should be stored separately adjacent to the mixer. Stockpiles should ideally be on a hard, concrete base, laid so that water will drain away, and separated into bays by division walls.

Stockpiles should be sited away from trees to prevent leaf contamination and kept free from general site and canteen rubbish.

Tarpaulins or plastic covers can be used to protect stockpiles from leaves, rubbish and rainwater.

In severe winter conditions the use of insulating blankets is to be recommended, to provide protection from frost and snow.

Timber

Timber is sawn or planed wood in its natural state – softwoods from coniferous trees and hardwoods from broadleaf trees. In general, softwoods are less decorative and tend to be used for structural work, painted joinery and trim. Hardwoods are more often used for decorative work, polished joinery and trim. Timbers readily absorb and lose moisture to achieve a balance with their surroundings. However this causes the timber to expand and shrink, which can cause it to distort, split and crack. In addition, damp and wet timber is highly susceptible to fungal decay.

Timber may be either supplied loose in individual lengths, in banded packs or in shrink-wrapped plastic packs.

Individual lengths should be off-loaded manually; long lengths and large sections may require a person at each end. Banded or wrapped loads are best off-loaded mechanically using the lorry-mounted device, a fork-lift truck or a crane.

Timber supplied in shrink-wrapped plastic packs should be stored in them until required for use. Care must be taken not to damage the plastic.

Carcassing timber and external joinery should be stored level horizontally and clear of the ground on bearers to prevent distortion and absorption of ground moisture. Piling sticks or cross-bearers should be placed between each layer, at centres of about 600 mm, to provide support and allow air circulation.

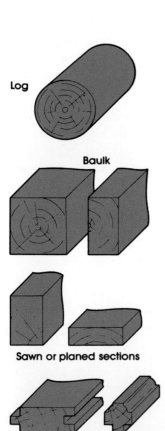

Figure 4.23 *Timber sections*

Figure 4.24 *Handling timber*

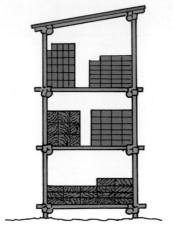

Figure 4.25 *Timber storage*

Figure 4.26
Trussed rafter storage

To protect timber against rain, frost, direct sunlight and atmospheric pollution, all stacks should be covered with a tarpaulin or polythene sheets, weighted or tied at the bottom. Care must be taken to allow free air circulation through the stack, to reduce the risk of fungal attack and condensation problems.

Internal trim and other planed sections may be stored horizontally in open-ended covered racks. Priming or sealing of trim and planed sections should be carried out on receipt, if it has not been done prior to delivery.

Trussed rafters are either supplied singly or in banded sets. They should be racked upright against a firm support, on eaves bearers. Alternatively, trussed rafters may be stored horizontally on close-spaced bearers, to give level support and prevent deformation. Stacks should be covered with a tarpaulin or polythene sheets, weighted or tied at the bottom. Care must be taken to allow free air circulation through the stack, to reduce the risk of fungal attack, condensation or connector corrosion problems.

Materials *Chapter 4*

Storage of hazardous products

Paint

This is a thin decorative and/or protective coating, which is applied in a liquid or plastic form and later dries out or hardens to a solid film covering a surface. Paints consist of a film former, known as the vehicle; a thinner or solvent (water, white spirit or methylated spirit, etc.) to make the coating liquid enough; and a pigment suspended in the vehicle to provide covering power and colour. Paint schemes require either the application by brush, spray or roller, of one or more coats of the same material (varnish, emulsion and solvent paints) or a build-up of different successive coats, each having their own functions (primer, undercoat and finishing coat).

162 **A Building Craft Foundation**

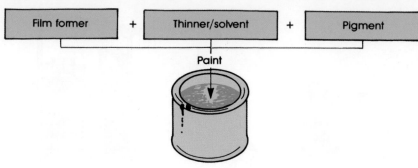

Figure 4.27

Paint formulation

◆ *Varnish* – paint without a pigment, used for clear finishing.
◆ *Emulsion* – a water-thinned paint for use on walls and ceilings.
◆ *Solvent paint* – based on rubber, bitumen or coal tar and used for protecting metals and water-proofing concrete, etc.
◆ *Primer* – may form a protective coat against moisture and corrosion, or act as a barrier between dissimilar materials. Also provides a good surface for bonding subsequent coats.
◆ *Undercoat* – a paint used on primed surfaces to give it a uniform body and colour on which a finishing coat can be successfully applied.
◆ *Finishing coat* – seals the surface, gives the final colour and provides the desired surface finish (flat, eggshell, gloss).

Paints are mainly supplied in 1-, 2.5- and 5-litre containers, and more rarely in bulk or trade 25-litre containers. These should be stored on shelves, in a secure store, at an even temperature. Each shelf and container should be marked with its contents.

Large containers should be placed on lower shelves to avoid necessary lifting.

Highly flammable liquids

did you know?

'Inflammable' and 'flammable' mean the same thing; 'liable to catch fire', or 'easily ignited', because it contains 'in' it is often taken to mean the opposite; 'non-flammable' means 'cannot be ignited'.

White spirit, cellulose, most special paint thinners and some formwork release agents give off a vapour that can ignite at room temperatures. The use of these liquids is controlled by the Highly Flammable Liquids and Liquefied Petroleum Gases Regulations. Up to 50 litres may be stored in a normal store, but because of the increased fire risk, larger quantities must be stored in special fire-resistant stores, normally placed at least 4 metres away from buildings, materials, workplaces and boundary fences.

All storerooms used for hazardous products should be no smoking areas; signs stating 'NO SMOKING – FLAMMABLE' and 'HIGHLY FLAMMABLE SUBSTANCES' must be prominently displayed.

Suitable fire extinguishers must be available to deal with the potential hazard. See chapter 2, 'Health and Safety'.

Rags used to mop up a spillage must **never** be left in the store. Rolled up dirty rags start to generate heat and can eventually burst into flames (spontaneous combustion). This creates an ignition source for the entire store contents.

Hazardous products may be used directly from their container or a smaller amount decanted into a 'kettle' for convenience. The surplus should be

Figure 4.28 *LPG cylinders*

Figure 4.29

Always store cylinders in an upright position

Figure 4.30

Ensure valves are closed

returned to the main container after use. **Never** store any substance in an unlabelled container.

Decanting

Consult the manufacturer's instructions on the container prior to decanting as certain substances **must not** be used or stored in another container for reasons of safety (it parts them from their use/safety instructions and renders the contents unknown).

To decant from container to kettle: dust the top of container; remove the lid with an opener; thoroughly stir the contents with a mixing knife to achieve an even consistency; pour the required quantity into the kettle, from the side opposite the manufacturer's instructions; use a brush to mop up the surplus on the rim or side of the container; scrape the brush on the edge of the kettle to remove surplus; firmly replace the container lid to prevent vapours escaping and dust and dirt getting in.

Decanting of hazardous products should be carried out in well-ventilated conditions.

Gases

Liquefied petroleum gas (LPG) sold commercially as propane and butane is supplied in pressurized metal cylinders ranging from 4 kg to 50 kg in size.

Cylinders should be stored upright in a well-ventilated fire-resisting storeroom, or in a secure compound away from any heat source. A minimum of two exits are normally required for an LPG store.

A sign stating 'NO SMOKING, HIGHLY FLAMMABLE SUBSTANCES' should be displayed. Empty and full cylinders should be stored separately and empty ones clearly marked as such.

Never use or store cylinders on their sides, nor near to or in excavations, in confined spaces or in other areas with restricted ventilation.

All valves should be fully closed. LPG is heavier than air and it will collect at low levels where an ignition source (discarded match, sparking power tool, dirty rags) could ignite it.

Storage of fragile or perishable materials

Bagged materials

Cement

Cement is manufactured from chalk or limestone and clay, which are ground into a powder, mixed together and fired in a kiln causing a chemical reaction. On leaving the kiln, the resulting material is ground to a fine powder. Hence a popular site term for cement is 'dust'. When water is added to the cement, another reaction takes place causing it to gradually stiffen, harden and develop strength.

Materials Chapter 4

◆ *Ordinary Portland cement (OPC)* – when hardened its appearance resembles Portland stone.
◆ *Rapid hardening Portland cement (RHPC)* – for cold weather use.
◆ *Sulphate-resisting Portland cement (SRPC)* – for use underground in high sulphate conditions.
◆ *White or coloured Portland cement* – made using white china clay; pigments are added for coloured cements.
◆ *High alumina cement (HAC)* – uses bauxite (aluminium oxide) instead of clay. It develops very early strength, which is much higher than OPC, although in the long term it has been found unstable and thus is now rarely favoured for structural work.

Cement is used in all forms of *in situ* and pre-cast concrete products, cement mortar, cement screeds and rendering.

Plaster

Plaster is applied on internal walls and ceilings to provide a jointless, smooth, easily decorated surface. External plastering is normally called rendering. Plaster is a mixture that hardens after application; it is based on a binder (gypsum, cement or lime) and water with or without the addition of aggregates. Depending on the background (surface being plastered) plastering schemes may require the application of either one coat, or undercoats to build up a level surface followed by a finishing coat.

◆ *Gypsum plaster* – for internal use, different grades of gypsum plaster are used according to the surface and coat. For undercoats, browning is generally used for brick and blockwork or bonding for concrete; for finishing coats, finish is used on an undercoat or board finish for plasterboard.
◆ *Cement–sand plaster* – used for external rendering, internal undercoats and water-resisting finishing coats.
◆ *Lime–sand plaster* – used for both undercoats and (rarely) finishing coats, although lime can be added to other plasters to improve their workability.

Lime

Lime is ground, powdered white limestone added to plaster and mortar mixes to improve workability.

Bagged material storage

Bagged materials may be supplied loose in individual bags or in unit loads shrink-wrapped in plastic on timber pallets.

Individual bags should be off-loaded manually, with the bag over the shoulder as the preferred method. Palleted loads are best off-loaded mechanically using the lorry-mounted device, a fork-lift truck or a crane.

Bags supplied in shrink-wrapped loads are best stored in these until required for use. Care must be taken not to damage the plastic.

Bags should be stored in ventilated, waterproof sheds, on a sound dry floor, with different products having their own shed to avoid confusion.

Bags should be stored clear of the walls and no more than eight to ten bags high. This is to prevent bags becoming damp through a defect in the outside wall, causing the contents to set in the bag. It also reduces the risk of compaction ('warehouse setting') of the lower bags due to the excessive weight of the bags above.

Bags should be used in the same order as they were delivered, known as 'first in, first out' (FIFO). This is to minimize the storage time and prevent the bag contents becoming stale or 'air setting'.

Where small numbers of bags are stored and a shed is not available, they may have to be stored in the open. Stack no more than six to eight bags high on timber pallets and cover with tarpaulins or polythene sheets weighted or tied at ground level.

Figure 4.31 *Handling bagged materials*

Figure 4.32 *Storage of bagged materials in different shed*

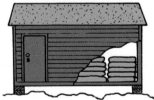

Figure 4.33 *Open storage of bagged materials*

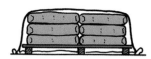

Plywood

Blockboard

Laminboard

Particle board (chipboard)

Plasterboard

Figure 4.34 *Types of sheet material*

Sheet materials

Supplied either as individual sheets, taped face to face in pairs (plasterboard), in banded bundles or unit loads on timber pallets; they may also be supplied in shrink-wrapped plastic packs. Various types of sheet material are illustrated in Figure 4.34.

Plywoods

Usually consist of an odd number of thin layers glued together with their grains alternating, for strength and stability. Used for flooring, formwork, panelling, sheathing and cabinet construction.

Laminated boards

Consist of strips of timber that are glued together, sandwiched between two plywood veneers. They are used for panelling, doors and cabinet construction.

Particle boards

Either chipboard (small chips and flakes) or waferboard (large flakes or wafers); both are manufactured using wood chips and/or flakes impregnated with an adhesive. They are used for flooring, furniture and cabinet construction.

Fibreboards

Made from pulped wood, mixed with an adhesive and pressed forming hardboard, medium board, medium density fibreboard (MDF) and insulation board. They are used for floor, wall, ceiling and formwork linings, insulation, display boards, furniture and cabinet construction.

Woodwool slabs

Made from wood shavings coated with a cement slurry; used for roof decks and as a permanent formwork lining.

Plastic laminate

This is made from layers of paper impregnated with an adhesive; used for worktops and other horizontal and vertical surfaces requiring decorative hygienic and hard-wearing properties.

Plasterboard

This comprises a gypsum plaster core sandwiched between sheets of heavy paper; used for wall and ceiling linings.

Materials **Chapter 4**

Sheet materials handling and storage

Individual sheets should be off-loaded manually. To avoid damage they should be carried on edge, which may require a person at each end. Banded bundles or palleted loads are best off-loaded mechanically using the lorry-mounted device, a fork-lift truck or a crane.

Sheets supplied in shrink-wrapped plastic packs should be stored in them until required for use. Care must be taken not to damage the plastic.

All sheet materials should preferably be stored in a warm, dry place; ideally stacked flat on timber cross-bearers, spaced close enough together to prevent sagging. Alternatively, where space is limited, sheet material can be stored on edge in a purpose-made rack, which allows the sheets to rest against the back board in a true plane.

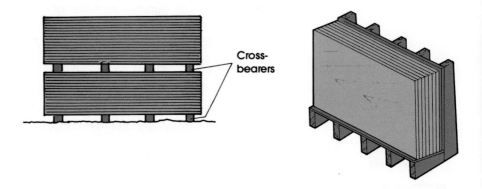

Cross-bearers

Figure 4.35

Stack of sheet material

Leaning sheets against walls on edge or end is not recommended as they will take on a bow, which is difficult to reverse.

Veneered or other finished surfaced sheets should be stored good face to good face, to minimize the risk of surface scratching.

Glass

Glass is a mixture of sand, soda, ash, limestone and dolomite that is heated in a furnace to produce molten glass. On cooling the molten mixture becomes hard and clear. Glass is supplied individually in single sheets or in banded timber packs. Sealed units and cut sizes may be supplied in shrink-wrapped plastic packs for specific purposes.

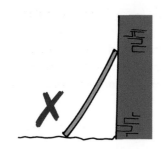

Figure 4.36 *Leaning of sheet material not recommended*

- ◆ *Drawn glass* – molten glass is drawn up between rollers in a continuous flow, cooled in water towers and cut into sheets. Patterned rollers may be introduced to create rough cast and patterned glass. Wire can be incorporated in the glass during the drawing to form wired glass, used for fire-resistant purposes. The surfaces of drawn glass are not perfectly flat, so when you look through your view is distorted. This means that large sheets of glass for shop fronts etc. have to be ground and polished perfectly flat to give undistorted vision. This type of glass is known as **polished plate**.
- ◆ *Float glass* – molten glass is floated on to the surface of liquid tin, and subsequently allowed to cool. When looked through it gives an undistorted view without the need for polishing.

◆ *Safety glazing* – 'at risk' areas of glazing such as fully glazed doors, patio doors, side panels and other large glazed areas, should contain a safety glazing material. Toughened safety glass is up to five times stronger than standard glass. If broken, it will break into fairly small pieces with dulled edges. Laminated safety glass is a sandwich of two or more sheets of glass interlaid with a plastic film. In the event of an impact the plastic holds the sheets of glass together. Depending on the number of layers, impacts from hammer blows and even gun shots can be resisted.

Storing glass

Glass should be stored in dry, wind-free conditions. Never store glass flat as it will distort and break. Sheets should be stood on one long edge almost upright at an angle of about 85 degrees. Timber and/or felt blocks should be used to prevent the glass coming into contact with rough surfaces, which can result in scratches or chips. Always stack sheets closely together and never leave spaces, as again this can lead to distortion and subsequent breakage.

Handling glass

When handling glass use laps to protect your palms. In addition, gauntlets may be worn to protect your lower arms and wrists.

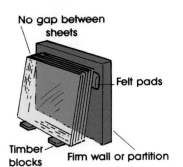

Figure 4.37 *Stacking glass*

Figure 4.38 *Use laps or gauntlets when handling glass*

Glass should be held firmly, but not too tightly as it may break. To balance the glass correctly: small panes should be carried under one arm and steadied on the front edge by the other hand; larger panes are held towards your body with one hand under the bottom edge and the other steadying the front; large sheets will require two people to handle them, walking in step, one on each side.

To prevent large panes and sheets 'whipping' when handling it is safer to carry two or more at a time.

Figure 4.39

Handling glass

Materials Chapter 4

Figure 4.40

Take a wide path at corners when moving lass

Figure 4.41

Handling chinaware

Ensure your route is clear of obstructions, keep about a metre out from the wall or building and take a wide path at corners. Do not stop or step back suddenly, since serious injury can be caused by a collision with another person.

If the pane of glass you are carrying breaks or slips, step clear and let it fall freely. Never attempt to catch it.

Vitreous chinaware

This is a ceramic material consisting of a mixture of sand and clay, which has been shaped, dried and kiln fired, to produce a smooth, hard, glassy surface material. Different colours are achieved by coating items with a prepared glaze solution before firing.

Used for sanitary appliances, such as WC (water closet) pans, WWP (waste water preventer) cisterns, wash basins and shower trays. They are supplied in various ways: as individual items, shrink-wrapped in plastic or with corners taped for protection; in unit loads shrink-wrapped on timber pallets or, increasingly, in bathroom sets including bath, shrink-wrapped on a timber pallet.

Vitreous chinaware handling and storing

Individual items should be off-loaded manually, with the item firmly gripped with both arms. Palleted loads are best off-loaded mechanically using the lorry-mounted device, a fork-lift truck or a crane.

Items are supplied in shrink-wrapped plastic loads and are best stored in them until required for use.

Individual items should be nested together on timber bearers, or alternatively stored in a racking system.

Figure 4.42

Storage of chinaware

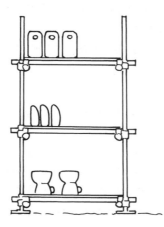

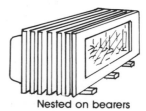

Nested on bearers

Storage of miscellaneous materials

Figure 4.43

Handling rolled materials

Figure 4.44

Rolled materials should be stored on end

Rolled materials

- ◆ *Bitumen* – either occurs naturally or distilled from petroleum, used for roofing felt and damp-proof courses (DPCs).
- ◆ *Metal* – mainly non-ferrous (not containing iron) such as copper, lead and zinc. Used for roof coverings, DPCs and flashing.
- ◆ *Plastic* – polythene, a thermoplastic (is softened by heat or solvent), used for DPCs and damp-proof membranes (DPMs).

Supplied as individual rolls, in banded bundles or unit loads shrink-wrapped on timber pallets.

Individual rolls should be off-loaded manually, with the roll over the shoulder as the preferred method. Palleted loads are best off-loaded mechanically using the lorry-mounted device, a fork-lift truck or a crane.

Rolls are supplied in shrink-wrapped plastic loads and are best stored in them until required for use.

All rolls should be stacked vertically on end, on a level, dry surface. Alternatively they may be stored in a racking system. However, again they should be vertical to prevent them rolling off and to reduce the risk of compression damage (e.g. the layers of bitumen rolls adhere together under pressure) due to excessive loads (this occurs if rolls are stacked horizontally on top of each other).

Ironmongery

Carpenters' locks, bolts, handles, screws and nails, etc. are 'desirable' items, which are most likely to 'walk' unless stored securely under the control of a storeperson.

They are supplied either as individual items, plastic bubble-packed on cards or in boxed sets, by amount, e.g. 10 pairs of hinges, 200 screws or 25 kg of nails, etc.

Large ironmongery deliveries may be supplied in mixed unit loads shrink-wrapped on timber pallets.

Individual items should be off-loaded manually and transferred immediately to the store. Palleted loads are best off-loaded mechanically using the lorry-mounted device, a fork-lift truck or a crane.

Large or heavy items should be stored on lower shelves to avoid unnecessary lifting.

Joinery

Doors, frames and units are supplied as individual items, in banded bundles, boxed, flat-packed or as ready assembled units and in unit loads on timber pallets. They may also be supplied in shrink-wrapped plastic packs for protection.

Figure 4.45

Handling frame joinery

Individual joinery items should be off-loaded manually. To avoid damage and stress they should be carried on edge or in their plane of use. This may require a person at each end. Banded bundles or palleted loads are best off-loaded mechanically using the lorry-mounted device, a fork-lift truck or a crane.

Items supplied in shrink-wrapped packs should be stored in them until required for use. Care must be taken not to damage the plastic.

All joinery items should preferably be stored in a warm, dry place; ideally stacked flat to prevent twisting and on timber cross-bearers, spaced close enough together to prevent sagging. Also see 'Timber' on page 158.

Leaning items against walls on edge or end is not recommended, as they will take on a bow, rendering them unusable.

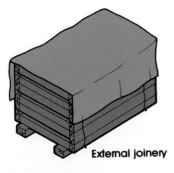
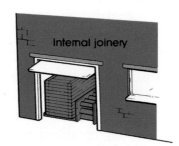

Figure 4.46 *Stacking and storage of frame joinery*

Figure 4.47 *Learning of frame joinery is not recommended*

Disposal of waste

Environmental awareness

Environmental awareness for all involved in construction activities means having an understanding of the need to conserve energy, prevent the wastage of materials and the proper disposal of waste. The government has promoted a number of initiatives, which are intended to conserve energy and reduce waste.

did you know?

In law, a 'duty of care' is a legal obligation imposed on individuals, requiring them to exercise a reasonable standard of care when undertaking tasks or acts that have the potential to harm others

- ◆ Building Regulations have been updated and improved to ensure that buildings are more energy efficient and building materials are more effectively used.
- ◆ The Environment Protection Act places a 'Duty of Care' on persons concerned with waste to:
  - ▶ Prevent the unauthorized or harmful disposal of waste
  - ▶ Prevent the escape of waste
  - ▶ Ensure waste is only transferred to authorized persons.

Waste classification

Waste is defined as any substance or objects that you discard, intend to discard or are required to discard, whether in a solid, liquid or gaseous form. Waste can be further classified as either:

- ◆ *Controlled waste* – all commercial and industrial waste other than that which is classified as hazardous
- ◆ *Hazardous waste* – all substances classified, marked or known to be Explosive, Oxidizing, Flammable, Toxic, Harmful, Irritant, Corrosive, Dangerous to the Environment, Poisonous or Carcinogenic etc.

Waste disposal

The construction industry in the UK has a major role to play in improving the quality of the environment. Traditionally the bulk of waste has been disposed of in landfill sites. However, the spiralling costs of waste disposal, along with government and European pressure, means that waste management on construction sites has to change.

To minimize the amount of waste all concerned must give priority to the following points in the order stated:

- ◆ **Eliminate** waste wherever possible:
  - ▸ Ovoid over-ordering of materials
  - ▸ Order in the lengths required
  - ▸ Arrange suitable storage areas to avoid damage and loss.
- ◆ **Reduce** the amount of waste created:
  - ▸ Keep all materials in their protective packaging until they are required for use to protect them from damage
  - ▸ Avoid as much double handling as possible, to result in less effort, damage and waste
  - ▸ Return unused materials back into storage, to aid site safety as well as reducing waste
  - ▸ Keep significant off-cuts for later use
  - ▸ Check that old stock of a material has been fully used before starting on a new batch.
- ◆ **Reuse** materials that are potential waste:
  - ▸ Use off-cuts where possible, e.g. off-cuts of plasterboard for repairs; off-cuts of timber for pegs and profile boards etc.
  - ▸ Reuse materials as many times as possible until they are no longer fit for purpose, e.g. formwork and site hoardings etc.
  - ▸ Reuse materials for alternative purposes, e.g. broken bricks, stonework and tiles can be used as hardcore below concrete slabs.
- ◆ **Recycle** waste materials wherever possible:
  - ▸ Segregate waste on-site into different types (see Table 4.1).
  - ▸ Store segregated waste in clearly marked separate skips or containers for potential recycling
  - ▸ Bricks from demolition works can be cleaned and recycled
  - ▸ Timber from demolition works can be recycled into new products, e.g. reclaimed oak and pine is in particular demand for recycling into furniture
  - ▸ Timber that can't be reused can be recycled for use in the production of chipboard and medium density fibreboard (MDF)
  - ▸ Surplus concrete can be crushed and recycled for use as a concrete aggregate (RCA)

Materials | Chapter 4

- ► Broken glass can be recycled for use in brick manufacturing, grit blasting, cement and concrete, fibreglass insulation and the manufacture of new glass
- ► Plasterboard and other gypsum products can be recycled for use in the manufacturing process
- ► Scrap metals can be recycled for use in new materials
- ► Packaging materials can also be recycled for use in new materials.
- ◆ **Dispose** of the remaining waste to a landfill site.

These actions will minimize the environmental impact of waste disposal and at the same time help to improve site safety and reduce construction costs.

Waste segregation

The segregation of waste into different types can aid recycling, help to minimize the cost of waste disposal and reduce the environmental impact of waste.

Waste containers and skips should be labelled with the following standard signs to encourage and improve the segregation of waste.

Table 4.1

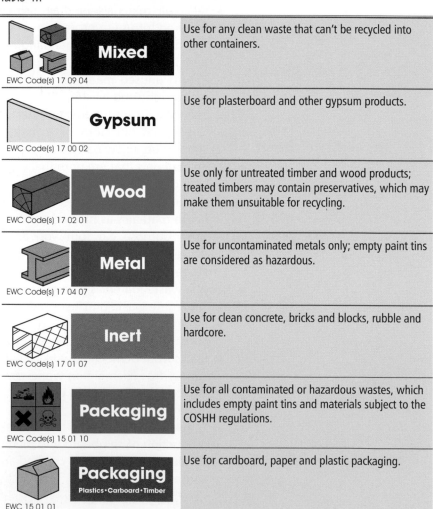

Mixed — EWC Code(s) 17 09 04	Use for any clean waste that can't be recycled into other containers.
Gypsum — EWC Code(s) 17 00 02	Use for plasterboard and other gypsum products.
Wood — EWC Code(s) 17 02 01	Use only for untreated timber and wood products; treated timbers may contain preservatives, which may make them unsuitable for recycling.
Metal — EWC Code(s) 17 04 07	Use for uncontaminated metals only; empty paint tins are considered as hazardous.
Inert — EWC Code(s) 17 01 07	Use for clean concrete, bricks and blocks, rubble and hardcore.
Packaging — EWC Code(s) 15 01 10	Use for all contaminated or hazardous wastes, which includes empty paint tins and materials subject to the COSHH regulations.
Packaging Plastics · Carboard · Timber — EWC 15 01 01	Use for cardboard, paper and plastic packaging.

Table 4.2 *Do's and Don'ts of waste disposal*

Do ✓	Don't ✗
Keep site tidy, collect waste regularly; a tidy site is a safer site.	Don't burn or bury waste on-site; it is illegal.
Place all waste in the appropriate container.	Don't overfill skips or containers.
Use enclosed containers or covered skips to prevent waste being blown about.	Don't mix hazardous waste with other controlled waste; it is illegal.
Seek advice from the site waste coordinator if you are unsure about the correct segregation of waste.	Don't contaminate one type of waste with another.
	Don't put liquid, flammable or hazardous wastes in with mixed waste.

activity

You are to assist the general foreman in the site planning of a small estate of four-bedroom houses and two-bedroom bungalows.

1. Photocopy or download the estate plan (Figure 4.48) and indicate on it the suggested positioning of the following:

   a) site office
   b) WC/washroom
   c) eating and changing facilities
   d) secure stores and compound.

2. State the facilities required on-site for storing the following:

   a) cement
   b) timber joists
   c) paint
   d) copper pipe
   e) joinery
   f) plumbing and electrical fittings.

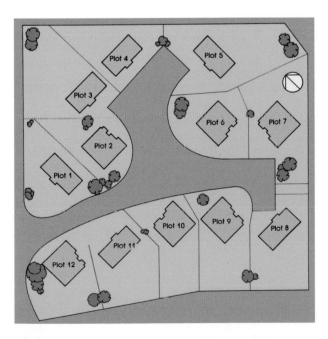

Figure 4.48

Estate plan

3. Name a bulk building material and state the reason why it should be stored clear of the ground.

4. State the purpose of covering stored building materials.

5. Explain why piling sticks or cross-bearers are used when stacking timber.

6. State why it is not good practice to store liquids in unmarked containers.

7. Explain the term 'first in, first out' when applied to the use of bagged materials.

8. State the reasons why rolled materials are normally stored vertically on end.

9. Describe how cut panes of glass should be stored prior to use.

10. Explain why flat storage is recommended for doors, frames and sheet materials.

11. Name an item of safety clothing used when handling glass.

12. Identify the materials from the following descriptions:
   a) a walling unit component having a standard format size including a 10-mm mortar allowance of 225 mm × 112.5 mm × 75 mm
   b) a paint used to form a protective coat against moisture and corrosion, or act as a barrier between dissimilar materials
   c) a sheet material that is formed by being floated on to the surface of liquid tin.

13. Name three types of hazardous waste.

14. Define what is meant by 'environmental awareness'.

15. In what skip or container should empty paint tins be placed?

Numerical Skills

This chapter is intended to provide the new entrant with an overview of the numerical skills that are required in the construction industry. Although its content is not assessed directly, knowledge of its contents is required in other assessed units. It is concerned with the types of calculations that you may be required to undertake on a day-to-day basis.

In this chapter you will cover the following range of topics:

◆ The number system
◆ Types of numbers
◆ Basic rules for numbers
◆ Units of measurement
◆ Applied mathematics
◆ Statistics
◆ Powers and roots of numbers
◆ Angles and lines
◆ Shapes and solids
◆ Formulae
◆ Measuring and costing materials.

The use of numbers is a communication skill widely required in the construction industry. Typical questions are:

◆ How many do we need?
◆ How long is it?
◆ What is the area or volume?
◆ How long will it take?
◆ How much will it cost?
◆ How much pay will I get?

You will need to revise all you know about numbers and how to use them for expressing answers to the above questions.

The number system

The figures or digits 0, 1, 2, 3, 4, 5, 6, 7, 8 and 9, and their position or place, are used to give the value of a number. We say that 264 is:

◆ two hundred and sixty-four, or
◆ 200 and 60 and 4, or
◆ 2 hundreds, 6 tens and 4 units.

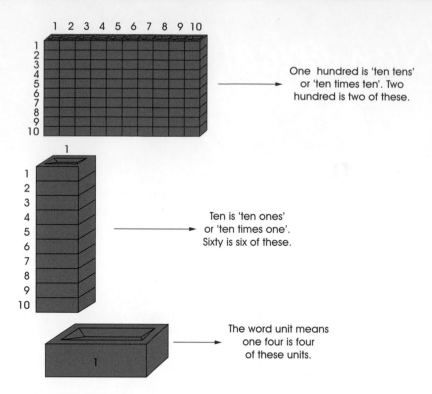

One hundred is 'ten tens' or 'ten times ten'. Two hundred is two of these.

Ten is 'ten ones' or 'ten times one'. Sixty is six of these.

The word unit means one four is four of these units.

Figure 5.1

The number system

Each figure has a different value when put in a different place, e.g.:

◆ In 246 the 6 stands for 6 units
◆ In 462 the 6 stands for 6 tens
◆ In 624 the 6 stands for 6 hundreds.

Counting in ones or units

Figure	Word
0	zero
1	one
2	two
3	three
4	four
5	five
6	six
7	seven
8	eight
9	nine
10	ten
11	eleven
12	twelve
13	thirteen
14	fourteen
15	fifteen

16	sixteen
17	seventeen
18	eighteen
19	nineteen
20	twenty
21	twenty-one
22	twenty-two
23	twenty-three
24	twenty-four
25	twenty-five

A zero means 'nothing' and is used in the number system to fill an empty space or place, e.g.:

◆ 102 is one hundred and two or one hundred, zero tens and two units. The zero fills the empty space in the place for the missing tens
◆ 120 is one hundred and twenty or one hundred, two tens and zero units. The zero keeps a place for the missing units.

Counting in tens

10	ten	=	1 ten
20	twenty	=	2 tens
30	thirty	=	3 tens
40	forty	=	4 tens
50	fifty	=	5 tens
60	sixty	=	6 tens
70	seventy	=	7 tens
80	eighty	=	8 tens
90	ninety	=	9 tens
100	one hundred	=	10 tens

Counting in large numbers

For numbers larger than 999 (nine hundred and ninety-nine), thousands are used (10 hundreds):

1000	one thousand = ten hundreds
2000	two thousand = twenty hundreds
10 000	ten thousand
20 000	twenty thousand
100 000	one hundred thousand
200 000	two hundred thousand
4240	four thousand, two hundred and forty
21 361	twenty-one thousand, three hundred and sixty-one
320 636	three hundred and twenty thousand, six hundred and thirty-six

Numerical skills

Chapter 5

did you know?

When using large numbers containing more than four digits, a thin gap or comma is normally added between each group of three figures starting from the right-hand end (units).

For numbers larger than 999 999, **millions** are used (1000 thousand):

1 000 000	one million = one thousand thousand
10 000 000	ten million
100 000 000	one hundred million
6 071 324	six million, seventy-one thousand, three hundred and twenty- four
24 650 150	twenty-four million, six hundred and fifty thousand, one hundred and fifty
502 081 015	five hundred and two million, eighty-one thousand and fifteen

For numbers larger than 999 999 999, **billions** are used (1000 million):

1 000 000 000	one billion = one thousand million
10 000 000 000	ten billion
100 000 000 000	one hundred billion

Number groups

Specific names are sometimes used for certain groups or quantities of numbers:

- ◆ 1 dozen = 12
- ◆ 1 gross = 144 (1 dozen dozen)
- ◆ 1 quire = 25 (sheets of paper)
- ◆ 1 ream = 500 (20 quire).

Roman numbers

Roman numbers are rarely used today, apart from some clock faces and some dates. They would be much harder to work with, as there is no zero to keep place value.

Roman number	Value
I or i	1
II or ii	2
III or iii	3
IV or iv	4
V or v	5
VI or vi	6
VII or vii	7
VIII or viii	8
IX or ix	9
X or x	10
L	50
C	100
D	500
M	1000

Roman numbers are written by putting the letters side by side:

MCXV	1115
MDCCLVI	1756
MMCCCLXIII	2363

Working on a photocopy, fill in the missing figures and words in Table 5.1.

Numbers in words	Numbers in figures	Billions	Millions	Thousands	Hundres	Tens	Units
Sixty-four thousand and one				800			1
Sixty-four thousand and one	64 001						
	607 098 644						
		1		5	6	4	0
Three hundred and fifty thousand, six hundred and eight							
			9	10	5	0	4
	10 964 001						
Nine million, six hundred and fifty thousand, four hundred and sixty-nine							
			2	800	6	4	9
	58 048						
Eighty-eight thousand and eight							

Table 5.1

1. Write the following numbers in both normal figures and roman figures:
   a) Sixteen
   b) One hundred and forty-five
   c) Two thousand, five hundred and six
   d) Five hundred and ninety-seven

2. List the following numbers in size order, starting with the lowest:
   a) 34 700
   b) 7043
   c) 60 047
   d) 120 012
   e) 3 404 152
   f) MMMDCCLXV

3. Write the following numbers in words:
   a) 42 304
   b) 8 704 312
   c) 1 001
   d) CCCLXV.

Types of numbers

Positive and negative numbers

Positive numbers

These have a value greater than zero. They may be written either without any sign in front of them or with a plus (+) sign, e.g.:

10   240   +1147   +523

Negative numbers

These have a value less than zero. They are written with a minus (–) sign in front of them, e.g.:

–10   –45   –115   –6

Negative numbers may be used to show a temperature below freezing point or on a bank statement to show an overdrawn balance.

Directed numbers

These are numbers that may be positive or negative.

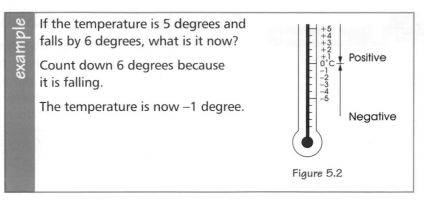

example

If the temperature is 5 degrees and falls by 6 degrees, what is it now?

Count down 6 degrees because it is falling.

The temperature is now −1 degree.

Figure 5.2

example

If the temperature was −2 degrees and rises by 5 degrees, what is it now?

Count up 5 degrees because it is rising.

The temperature is now 3 degrees (or +3 degrees).

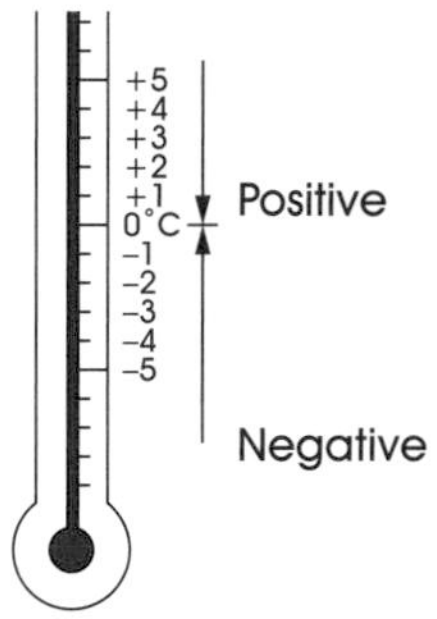

Figure 5.3

Fractions

Fractions are parts of a whole number, e.g. for half a brick we write the fraction:

- ◆ $\frac{1}{2}$ a brick, which means one part out of two parts.

For three-quarters of a plank of wood, we write the fraction:

- ◆ $\frac{3}{4}$ of a plank, which means 3 parts out of 4 parts.

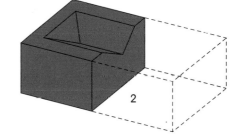

Figure 5.4 *Half*

- ◆ The top number of a fraction is the **numerator**, the bottom one is the **denominator**.
- ◆ A fraction like $\frac{1}{2}$ or $\frac{3}{4}$ is called a **proper fraction**.
- ◆ A fraction like $\frac{3}{2}$ or $\frac{4}{3}$ is called an **improper fraction**.
- ◆ A fraction like $1\frac{1}{2}$ or $2\frac{1}{4}$ is called a **mixed number**.

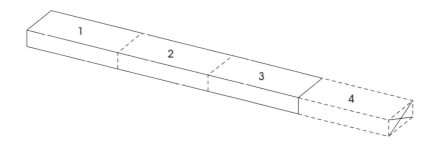

Figure 5.5

Three-quarters

Subtraction

Subtraction involves taking things away.

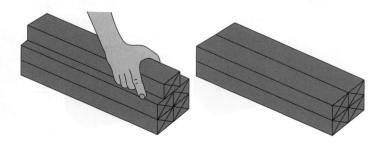

Figure 5.8

Subtraction

5 pieces of wood, take away (minus) 1 leaves 4
5 − 1 = 4

Again when subtracting, line up the units underneath each other and subtract the columns starting from the right.

example	Subtract 48 from 694
	8 1 6⁄94 48 − 646 **ANSWER:** <u>646</u>

◆ When subtracting the units, we say in this example '8 from 4 will not go, so borrow 1 from the tens columns', making 14 units and 8 tens.
◆ When subtracting the next column, we must remember there is one less at the top.

Subtraction of decimals

Numbers containing decimal points are again written down with the points lined up underneath each other.

example	16.697 − 8.565
	0 1 1⁄6.697 8.565 − 8.132 **ANSWER:** 8.132

Multiplication

This is a quick way of adding groups of equal numbers:

4 + 4 + 4 + 4 + 4 = 20 is the same as 4 × 5 = 20

You do not have to count the number of tiles on the wall to tell that there are 20.

There are 5 lots of 4 tiles. When using the quick way we say 'Four tiles multiplied by five lots equals 20 tiles' or simply 'Four by five equals 20'.

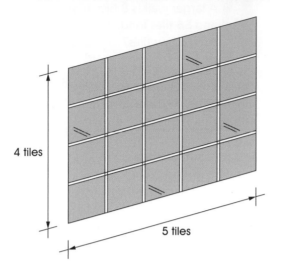

Figure 5.9

Multiplication

4 tiles

5 tiles

Hence:

$4 \times 5 = 20$

20 is called the **product** of 4 and 5. The product of any number up to 10 is shown in a **multiplication table**. In order to multiply you must know your multiplication tables for products at least up to 10×10. The example shown in the multiplication table is for the product of 4×5.

1	2	3	4	5	6	7	8	9	10
2	4	6	8	10	12	14	16	18	20
3	6	9	12	15	18	21	24	27	30
4	8	12	16	(20)	24	28	32	36	40
5	10	15	20	25	30	35	40	45	50
6	12	18	24	30	36	42	48	54	60
7	14	21	28	35	42	49	56	63	70
8	16	24	32	40	48	56	64	72	80
9	18	27	36	45	54	63	72	81	90
10	20	30	40	50	60	70	80	90	100

Figure 5.10

Multiplication table

To find 4×5, look at the intersection of row 4 and column 5. There you find the product of 4 and 5, which is 20.

Numerical skills   Chapter 5

Multiplication of large numbers

example

A larger wall is 8 tiles high
and 56 tiles long.
How many tiles?

Note:
8 × 56 means '8 lots of 56
make' or '8 by 56'

$$\begin{array}{r} 56 \\ 8 \times \\ \hline 448 \\ {\scriptstyle 4} \end{array}$$

ANSWER: 448 tiles

◆ Start with the units. We say in this example '8 × 6 is 48'. (Use the multiplication table.)
◆ Put the unit figure down (8) and carry the tens forward (4).
◆ Then continue to multiply 5 × 8, which is 40 plus the carried 4 makes 44.
◆ The product of 8 × 56 tiles is 448 tiles.

example

Multiply 146 × 92

$$\begin{array}{r} 146 \\ 92 \\ \hline {\scriptstyle 1} \\ 292 \times \\ {\scriptstyle 4\,5} \\ 13140 \\ \hline {\scriptstyle 1} \\ 13432 \end{array}$$

ANSWER: 13432

◆ In this example, first multiply 146 × 2, to give 292.
◆ Then multiply 146 × 90. Do this by putting a zero down in the units and multiply 146 by 9.
◆ 9 times 146 is 1314.
◆ Add the two rows to get 13 432.

Multiplication of decimals

When multiplying, any decimal points can be ignored until the two sets of numbers have been multiplied together.

example

11.6 × 4.5

Two places to the right of the decimal point to start with.

Count 2 places to left for point in answer.

$$\begin{array}{r} 116 \\ 45 \times \\ \hline {\scriptstyle 3} \\ 580 \\ {\scriptstyle 2} \\ 4640 \\ \hline {\scriptstyle 1\,1} \\ 5220 \\ {\scriptstyle 2\,1} \end{array}$$

ANSWER: 52.20

The position of the decimal point can be located by the following rule: the number of figures to the right of the decimal point in the answer will always equal the total number of figures to the right of the decimal point to start with.

Division

Division involves sharing or dividing things into equal parts. It is the opposite of multiplication.

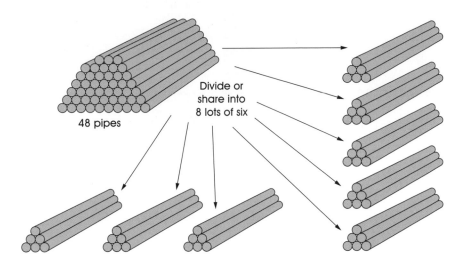

Divide or share into 8 lots of six

48 pipes

Figure 5.11

Division

Division using tables

Multiplication tables can be used backwards for division.

1	2	3	4	5	6	7	8	9	10
2	4	6	8	10	12	14	16	18	20
3	6	9	12	15	18	21	24	27	30
4	8	12	16	20	24	28	32	36	40
5	10	15	20	25	30	35	40	45	50
6	12	18	24	30	36	42	(48)	54	60
7	14	21	28	35	42	49	56	63	70
8	16	24	32	40	48	56	64	72	80
9	18	27	36	45	54	63	72	81	90
10	20	30	40	50	60	70	80	90	100

Figure 5.12

Multiplication table used for division

- ◆ Use the multiplication table. Look for 48 in the 6th row.
- ◆ Read up the column to get 8.
- ◆ Hence 48 shares into 8 equal lots of 6.
- ◆ We say '48 divided by (÷) 8 equals (−) 6'.

Division by calculation

Divide 384 by 6. This may be written down as 384 ÷ 6.

The 6 is called the **divisor**, 384 is called the **dividend** and the answer is called the **quotient**.

Note how the calculation is laid out in the example. For division we work from the left, not from the right, and the answer is written above, not below.

example

384 ÷ 6

```
          64
   6 | 384  ÷
       36
        24
```

ANSWER: 64

◆ First we try to find how many times 6 shares into 3.
◆ We say 6 into 3 will not go. So there is nothing to enter in the first position of the answer line.
◆ Then we try 6 into 38 (from table) and the nearest number below 38 in the 6th row is 36.
◆ Read up the column to get 6.
◆ Put the 6 above the 8 in the second position of the answer line.
◆ Subtract 36 from 38 to leave 2.
◆ 'Bring down' the 4 to make 24.
◆ Divide 24 by 6 (from table) equals 4. We say '6 into 24 goes 4 times'.
◆ Put the 4 next to the 6 in the third position of the answer line.

This gives the answer, so that 384 ÷ 6 = 64.

Your division can be checked by multiplying back. This check proves the working out is correct: 6 × 64 = 384.

example

6 × 64

```
       64
        6  ×
        2
      384
```

ANSWER: 384

Division with a remainder

When there is a remainder (left over at the end of the division) the operation can be continued by inserting a decimal point and 'bringing down' zeros. In the example:

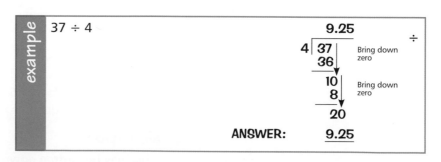

example

37 ÷ 4

```
           9.25              ÷
    4 | 37        Bring down
        36        zero
         10       Bring down
          8       zero
         20
```

ANSWER: 9.25

♦ We say '4 into 3 will not go'. (Nothing is entered in the first position of the answer line.)

♦ We say '4 into 37 goes 9 remainder 1'. (9 is entered into the second position of the answer line.)

♦ Insert decimal point next to 9 and bring down zero next to the remainder 1 making 10.

♦ We say '4 into 10 goes 2 remainder 2'. (2 is entered into the answer line after the decimal point.)

♦ Bring down another zero next to the remainder 2 making 20.

♦ We say '4 into 20 goes 5 and nothing left over'.

This gives the answer, so that, 37 ÷ 4 = 9.25.

Division of decimal numbers

When dividing numbers with decimal points, we need first to make the divisor a whole number. We do this by moving its decimal point a number of places to the right until it is a whole number, but to compensate we must also move the decimal point in the dividend by the same number of places. (We may have to add zeros to do this.)

example

164.6 ÷ 0.2

$$164.6 \div 0.2$$
$$1646 \div 2$$

Move both the decimal points until the division is a whole number.

$$\begin{array}{r} 823 \\ 2\overline{)1646} \end{array} \quad \div$$

ANSWER: 823

Operating with fractions

Fractions are best converted into decimals before proceeding with addition, subtraction, multiplication or division operations. Conversion is done by dividing the bottom number into the top number.

example

Express $\frac{7}{8}$ as a decimal.

$$\begin{array}{r} 0.875 \\ 8\overline{)7.0} \\ 64 \\ \hline 60 \\ 56 \\ \hline 40 \end{array}$$

ANSWER: 0.875

Combined operations

You will need to understand the following types of mathematical statements:

- 12 − (6 + 4)
- 3 × (4 − 1) usually written as 3(4 − 1)
- 3 × (12 ÷ 3) usually written as 3(12 ÷ 3)
- (40 ÷ 2) + (4 × 6)

Rules for combined operations

You must work out the operation contained in the brackets ( ) first before proceeding.

You must then do multiplication and division before addition and subtraction.

A useful made-up word 'BODMAS' can help you remember the order in which calculations should be undertaken:

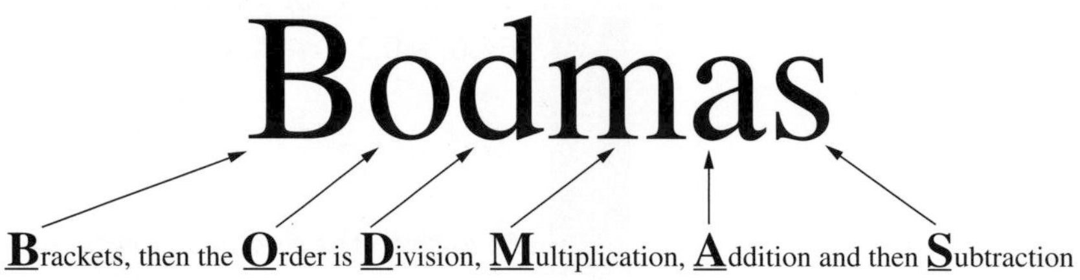

Bodmas

Brackets, then the **O**rder is **D**ivision, **M**ultiplication, **A**ddition and then **S**ubtraction

Figure 5.13

Order of undertaking a calculation

- So, 12 − (6 + 4) gives 12 − 10 = 2
- 3(4 − 1) gives 3 × 3 = 9

example

3(12 ÷ 3)

$$3(12 ÷ 3)$$
$$= 3(4)$$
$$= 3 × 4$$
$$= \underline{12}$$

40 ÷ 2 + (4 × 6)

$$40 ÷ 2 + (4 × 6)$$
$$= 40 ÷ 2 + (24)$$
$$= 20 + 24$$
$$= \underline{44}$$

6. 360 + 48 − 16 =

7. 480 ÷ 7 + 3 =

8. 12 460 + 750 − 192 =

9. 690 x 53 + 46 =

10. 74$^{1}/_{4}$ (3 + (15 −7)) =

Rough checks

Approximate answers

Common causes of incorrect answers to calculation problems are incomplete workings out and incorrectly placed decimal points. Rough checks of the expected size of an answer and the position of the decimal point would overcome this problem. These rough checks can be carried out quickly using approximate numbers:

◆ 4.65 × 2.05 ÷ 3.85
◆ For a rough check say 5 × 2 ÷ 4 = 2.5
◆ The actual correct answer is 2.476

The rough checks and the correct answer are of similar size. This confirms that the answer is 2.476 and not 0.2476 or 24.76 etc.

Rough checks will be nearer to the correct answer if, when choosing approximate numbers, some are increased and some are decreased. In cases where the rough check and the correct answer are not of the same size the calculation should be reworked to find the cause of the error.

Electronic calculations

All basic electronic calculators look similar. They include the following main key functions:

Numbered keys	0 1 2 3 4 5 6 7 8 9
Operation keys	+ − ÷ ×
Equals key	=
Decimal point key	·
Square root key	√
Percentage key	%
Clear last entry key	CE
Clear all entries key	AC or ON/C

Figure 5.14

Main key functions on a basic calculator

did you know?

Once you have an understanding of numbers a calculator may be used.

did you know?

For most modern calculators, the first operation of the = key may be omitted since the next key automatically proceeds with the calculation to give the desired result.

The operation of your calculator will vary depending on the model; therefore consult the booklet supplied with it before use. After some practice you should be able to operate your calculator quickly and accurately.

In the following example, first press the AC key to clear all numbers in the display. Then press the keys listed in the left-hand column in turn. Check that the correct numbers appear in the display.

example

$55.335 \times 2.1 \div 3.52$

Key	*Display*
AC or ON/C	0.
5 5 · 3 3 5	55.335
×	55.335
2 · 1	2.1
=	116.2035
÷	
3 · 5 2	3.52
=	33.012357

Answer 33.012357

The process of approximating answers covered on page 189 should be carried out even when using an electronic calculator, as wrong answers are often the result of miskeying; even a slight hesitation on a key can cause a number to be entered twice.

measuring up

Use your calculator to find the following values, round off your answers to 3 decimal places if required. Confirm your answer with a rough check.

11. Multiply 16×29

12. Find the product of 946, 18 and 46

13. Work out 347.2 times 3.1429

14. Find the quotient of $62.5 \div 2.4$

15. Work out 648.3 divided by 8.92

Rounding numbers

Number of decimal places

For most purposes, calculations that show three decimal figures are considered sufficiently accurate. These can therefore be rounded off to three decimal places. This however entails looking at the fourth decimal figure; if it is a five or above add one to the third decimal figure. Where it is below five ignore it.

> **example**
>
> 33.012357 becomes 33.012
>
> 2.747642 becomes 2.748

Number of significant figures

On occasions a number may have far too many figures before and after the decimal place for practical purposes. This is overcome by expressing it to 2, 3 or 4 significant figures (S.F. or sig. fig.).

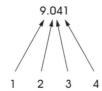

Figure 5.15

This illustration shows a decimal number expressed to 4 S.F.

> **example**
>
> Express 68.936102 to 4 S.F.
>
> ANSWER: **68.94**

In the example, we have applied the same rule as for rounding numbers to a number of decimal places. If the next figure after the last S.F. is 5 or more then round up the last S.F. by 1. See more examples below:

Number	To 3 S.F.	To 2 S.F.	To 1 S.F
6.308	6.31	6.3	6
5368	5370	5400	5000
0.051308	0.0513	0.051	0.05
0.1409	0.141	0.14	0.1

After rounding up to a number of significant figures, you must take care not to change the place value. For example, 4582 is 4600 to 2 S.F., so zeros are added to the end in order to maintain the place value.

However, 4.582 = 4.6 to 2 S.F., not 4.6000, since trailing zeros are not added after the decimal point.

measuring up

Chapter 5 Numerical skills

16. Complete the following table:

Number	To 3 S.F.	To 2 S.F.	To 1 S.F
5.874			
9643			
0.048739			
0.17973			

Units of measurement

Metrification of measurements in the construction industry is almost total. Previously imperial units of measurement were used. However, you may still come across them occasionally, so knowledge of both is required. To avoid the possibility of mistakes all imperial measurements should be converted to metric. All calculations and further work can then be undertaken in metric units of measurement.

The metric units used are also known as SI units (système international d'unités). For most quantities of measurement there is a base unit, a multiple unit and a sub-multiple unit. For example the base unit of length is the metre (m). Its multiple the kilometre (km) is a thousand times larger: m × 1000 = km. The sub-multiple unit of the metre is the millimetre (mm), which is a thousand times smaller: m ÷ 1000 = mm.

It is an easy process to change from one unit to another in the metric system. The decimal point is moved three places to the left when changing to a larger unit and three places to the right when changing to a smaller unit.

> **example**
>
> Change 6500 m to km
>
> 6 5 0 0   Move point 3 places
>
>   3 2 1   ANSWER: 6.5 km
>
> Change 0.55 m to mm
>
> 0 . 5 5 0   ANSWER: 550 mm
>
>   1 2 3

The units you will most frequently come across in the construction industry are covered in the following examples.

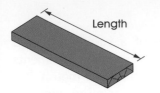

Figure 5.16

Length of plank in metres

Length

Length is a measure of how long something is from end to end. The length of this screw is 75 mm.

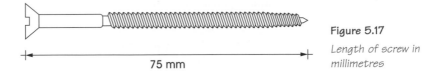

75 mm

Figure 5.17

Length of screw in millimetres

Long lengths

Lengths, such as the distance between two places on a map, are measured in kilometres (km). Miles are an imperial measure:

◆ 1 mile = 1760 yards
◆ 1 mile = 1609 km
◆ 1 km = 0.62 miles.

Figure 5.18

Long length in kilometres

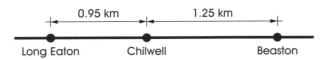

0.95 km   1.25 km

Long Eaton   Chilwell   Beaston

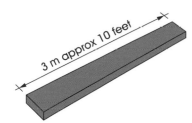

Figure 5.19

Metric/imperial length comparison in metres/feet

Intermediate lengths

Lengths, such as a piece of timber, are measured in metres (m). Feet (foot) and yards are imperial measurements:

◆ 3 feet = 1 yard
◆ 1 yard = 0.9144 metres
◆ 1 foot = 0.3048 metres
◆ 1 metre = 1.0936 yards
◆ 1 metre − 3.281 feet.

Small lengths

Lengths, such as a brick, are measured in millimetres. Inches are an imperial measure:

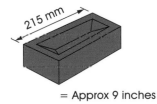

= Approx 9 inches

Figure 5.20

Metric/imperial length comparison in millimetres/inches

◆ 12 inches = 1 foot
◆ 1 inch = 25.4 millimetres.

Sometimes centimetres (cm) are encountered as an intermediate measure of length:

◆ 1 metre = 100 centimetres
◆ 1 centimetre = 10 millimetres.

However the centimetre is not used in the construction industry. Measurement should always be given in metres or millimetres, e.g. 55 cm is written as either 0.55 m or 550 mm.

Area

Area is concerned with the extent or measure of a surface. A patio 3 m × 4 m has an area of 12 square metres. Two linear measurements multiplied together give area (square measure). Both square metres and square millimetres are used for area. These are written as m² or mm². For example 12 square metres = 12 m².

- 1 m² = 1.196 square yards
- 1 m² = 10.764 square feet
- 1 mm² = 0.00155 square inches.

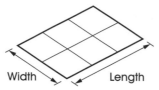

Figure 5.21 *Width × length = area*

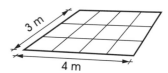

Figure 5.22 *Area = 3 × 4 = 12 m²*

Large areas

Large areas are measured in metric hectares or imperial acres.

- 1 hectare = 10 000 m²
- 1 hectare = 2.471 acres
- 1 acre = 4840 square yards.

Volume

Volume is concerned with the space taken up by a solid object. A room that is 4 metres wide by 5 metres long and 3 metres high has a volume of 60 cubic metres. The three linear measurements multiplied together give the volume (cubic measure). Cubic metres and cubic millimetres are used for volume. These are written as m³ or mm³, i.e. 60 cubic metres = 60 m³.

- 1 m³ = 1.31 cubic yards
- 1 m³ = 35.335 cubic feet
- 1 mm³ = 0.0000611 cubic inches.

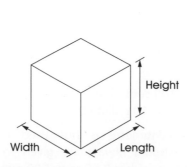

Width × Length × Height = Volume

Figure 5.23

Width × length × height = volume

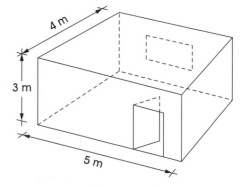

Volume = 4 × 5 × 3
= 60 m³

Figure 5.24

Volume = 4 × 5 × 3 = 60 m³

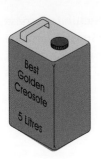

Figure 5.25 *Capacity*

Capacity

Capacity is concerned with the amount of space taken up by liquids or the amount of liquid that can fit into a given volume. Litre is the unit of capacity (to avoid confusion with the number 1 'litre' is best written in full) and millilitre (ml) is the sub-multiple: 1 litre = 1000 ml. Gallons, pints and fluid ounces are the imperial units of capacity.

◆ 1 gallon = 8 pints
◆ 1 pint = 20 fluid ounces
◆ 1 litre = 1.76 pints
◆ 1 litre = 0.22 gallons
◆ 1 ml = 0.035 fluid ounces
◆ 1 ml = 0.0017 pints.

Capacity is linked with volume. A carton 100 mm × 100 mm × 100 mm contains a litre of water. It would take 1000 of these cartons to fill 1 m³.

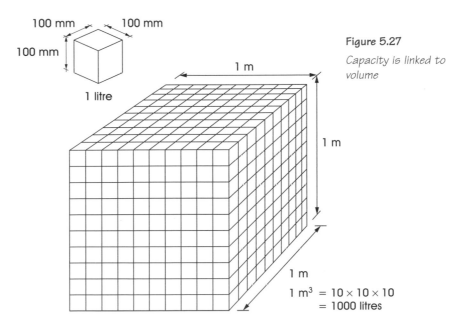

Figure 5.26

5 litres is more than 1 gallon;
4.5 litres = approx. 1 gallon

There are 1000 litres in one cubic metre.

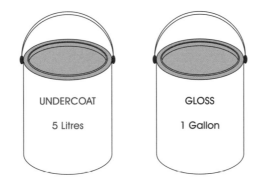

Figure 5.27

Capacity is linked to volume

$$1 \text{ m}^3 = 10 \times 10 \times 10$$
$$= 1000 \text{ litres}$$

Figure 5.28

1200 cc engine = 1.2 litre engine

Although not used in construction, 1 ml = 1 cubic centimetre (cc). Thus a car with a 1200 cc engine is 1200 ml or 1.2 litres in size.

Mass

Mass is concerned with the weight of an object. The metric unit of mass is the kilogram (kg). Its multiple is the tonne (to avoid confusion with the imperial ton no abbreviation is used). The sub-multiple unit is the gram (g); for very small objects the milligram (mg) is used. Imperial units are tons (ton), hundredweights (cwt), stones (st), pounds (lb) and ounces (oz).

- ◆ 1 tonne = 1000 kg
- ◆ 1 kg = 1000 g
- ◆ 1 g = 1000 mg
- ◆ 1 ton = 2240 lb
- ◆ 112 lb = 1 cwt
- ◆ 14 lb = 1 st
- ◆ 1 lb = 16 oz
- ◆ 1 tonne = 2205 lb
- ◆ 1 kg = 2.2 lb
- ◆ 30 g = 1 oz.

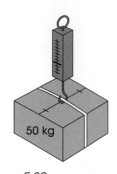

Figure 5.29

Mass is concerned with weight

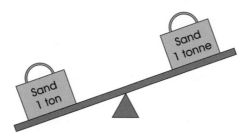

Figure 5.30

1 ton is heavier than 1 tonne

Density

This is a measure of a material's consistency. A more solid material is generally considered to be denser, but a true measure of density can only be calculated by measuring its weight and volume. The ratio of weight (mass) to volume is a measure of density, thus:

Mass ÷ Volume = Density

- ◆ The density of water is 1000 kg/m³
- ◆ The density of softwood is typically 450 kg/m³
- ◆ The density of concrete is typically 2400 kg/m³.

Note that to compare density of different materials, the ratio of weight to volume has to be converted to a standard unit of measurement – in this case kg/m³.

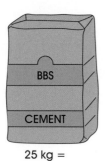

25 kg = approx 55 lb

Figure 5.31

Metric/imperial mass comparison in kilograms/pounds

Specific gravity

The density of materials is normally expressed in smaller units than kg/m³ by dividing values by 1000. This is referred to as the specific gravity (sg) and the specific gravity of water is the value 1 sg. Materials heavier than water have an sg greater than 1 sg and sink; materials lighter than water have an sg smaller than 1 sg and float.

Density ÷ 1000 = Specific gravity or relative density

◆ The specific gravity of water is 1 sg
◆ The specific gravity of softwood is typically 0.45 sg
◆ The specific gravity of concrete is typically 2.4 sg.

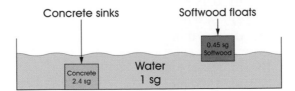

Figure 5.32

Specific gravity

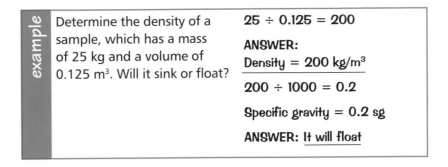

example	Determine the density of a sample, which has a mass of 25 kg and a volume of 0.125 m³. Will it sink or float?	25 ÷ 0.125 = 200 **ANSWER:** Density = 200 kg/m³ 200 ÷ 1000 = 0.2 Specific gravity = 0.2 sg **ANSWER:** <u>It will float</u>

Imperial to metric conversion

An approximation between imperial and metric measurements can be made using the comparisons given.

example	Convert 350 feet to metres 350 × 0.3048 = 106.68 ANS: <u>350 feet = 106.68 m</u> Convert 12 yards to metres 12 × 0.9144 = 10.9728 ANS: <u>12 yards = 10.973 m</u> Convert 1 gallon to litres 1 ÷ 0.22 = 4.545 ANS: <u>1 gallon = 4.545 litres</u> Convert 10 pounds to kilograms 10 ÷ 2.2 = 4.545 ANS: <u>10 pounds = 4.545 kg</u>

Numerical skills Chapter 5

17. Change to millimetres:
   a) 1.2 m
   b) 0.95 m
   c) 24.6 m
   d) 0.070 m

18. Express in metres:
   a) 1264 mm
   b) 920 mm
   c) 21 950 mm
   d) 68 mm

19. Convert 6 feet 6 inches into metres.

20. Change to kilograms:
   a) 1.2 tonne
   b) 5500 g
   c) 0.5 tonne
   d) 250 g

21. Convert 10 lb into kilograms.

22. You are asked to purchase 2 dozen 3-inch screws. The supplier only has metric screws in the following sizes: 50 mm, 62 mm, 75 mm, 82 mm and 100 mm. How many and what size will you buy?

23. You are informed that the last time a factory was painted out it took 10½ gallons of paint. How many 5-litre tins of paint are required this time?

24. You are told over the phone to pick up a 4 foot × 2 foot paving slab. Which is the nearest size metric slab?
   a) 450 × 900 m
   b) 600 × 900 mm
   c) 600 × 1200 mm
   d) 900 × 1200 mm

Time

This is a measure of the continued progress of existence, i.e. the past, the present and the future. The second is the main unit of time:

Figure 5.33

The time shown is: 3.55 or we can say 5 minutes to 4

- 60 seconds (s) = 1 minute (min)
- 60 minutes = 1 hour (h)
- 24 hours = 1 day
- 7 day = 1 week (wk)
- 31, 30, 29 or 28 days = 1 month (mth)

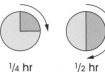

| ¼ hr | ½ hr | ¾ hr |
| 15 min | 30 min | 45 min |

Figure 5.34

Parts of an hour

did you know?

30 days for September, April, June and November. All the rest have 31 excepting February alone, which has 28 and in a leap year 29.

- 13 weeks = 1 quarter
- 26 weeks = 1 half year
- 12 months = 1 year
- 365 days = 1 year
- 366 days = 1 leap year (every 4 years to include 29 February)
- 52 weeks = 1 year

In the winter months we use GMT (Greenwich Mean Time). In the summer our clocks are put forward by 1 hour, called British Summer Time (BST).

Times in other parts of the world will not be the same due to time zone differences. The time of day is normally expressed using a 12-hour clock, with the morning and afternoon being distinguished by the use of AM or PM.

- AM is in the morning (ante meridian)
- PM is in the afternoon (post meridian)
- midday or noon is usually written as 12 AM
- midnight is usually written 12 PM.

Figure 5.35

24-hour clock

The 24-hour clock

Timetables are normally based on the 24-hour clock. This runs from 0000 hrs at midnight, to 2400 again at midnight.

- 8.30 AM is 0830 hours; 8.30 PM is 2030 hours
- Noon is 1200 hours; midnight is 0000 or 2400 hours
- 6.45 AM is 0645 hours; 6.45 PM is 1845 hours.

did you know?

Add 12 hours to the PM hours and drop the point to convert to the 24-hour clock.

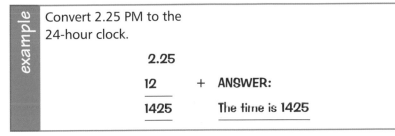

example

Convert 2.25 PM to the 24-hour clock.

```
  2.25
 12       +    ANSWER:
 ----
 1425          The time is 1425
```

Numerical skills

Chapter 5

Use the portion of a bus timetable to answer the following:

Long Eaton	Chilwell	Beaston	Nottingham
0700	0715	0730	0745
0730	⟶	⟶	0800
0800	0815	0830	0845
0830	0845	0900	0915
1000	⟶	1020	1040
1230	1245	1300	1315
1400	1415	⟶	1435

1. Which bus is the fastest between Long Eaton and Nottingham?

2. You arrive at Chilwell at 7.15 AM. How long is it before the next bus arrives?

3. How long does it take the fastest bus to travel between Chilwell and Nottingham?

4. You wish to arrive in Nottingham by 9 AM. Which bus from Long Eaton will you catch?

Money

Figure 5.36

UK notes and coins

Pounds (£) and pence (p) are the UK monetary units: £1 (pound) = 100p (pence).

£6.50 = 6 pounds 50 pence and is often read as 'six pounds fifty' and £9.07 is read as '9 pounds 7' or '9 pounds 7 pence'. It is important to remember the zeros as they keep the place for the missing units and tens.

Calculations with money can be simply undertaken using the previously covered rules of numbers.

example

Add £10.65 and £32.40

```
  10.65
  32.40  +
   ¹
  43.05
```

ANSWER:
£43.05

---

example

You give a £10 note to pay for a £6.47 lock. What is the change?

£10 – £6.47

```
  ¹ ¹
  10.00
  ⁷ ⁵
  6.47 –
  3.53
```

ANSWER:
There is **£3.53** change

---

example

Four people are to get an equal share of £145 as a bonus. How much will each receive?

£145 ÷ 4

ANSWER
Each will receive **£36.25**

```
      36.25
  4 | 145
      12
      25
      24
      10
       8
      20
```

Applied mathematics

Wages and salaries

People who work for an employer are paid either a wage or a salary. **Wages** are based on an hourly rate and normally paid weekly.

A basic week consists of a set number of hours, typically up to 40 hours. The amount paid per hour is known as the **basic rate**. After completing the basic week, additional work is paid at an **overtime rate**, time and a half or double time.

example

Suppose you are paid £7.50 an hour for the first 8 hours worked in a day and time and a half after that. How much will you earn for working a 10-hour day?

Basic 8-hour day = £7.50 × 8 = £60

2 hours overtime = £7.50 × 2 × 1.5 = £22.50

Total day's pay = £82.50

A **bonus** is sometimes paid by employers for doing more than a set amount of work in a week.

Hourly paid workers may have to fill in a **time sheet**, to show the hours they have worked.

Salaries are based on a fixed annual amount, which is normally paid monthly.

A project manager is paid a salary of £28 500 pa.

Monthly salary = £28 500 ÷ 12 = £2375

BBS Recruitment Solutions: WEEKLY TIME SHEET		Name: **J. JAMES**				Works No. **57**
Day	Date	Start Time	Lunch	Finish Time	Total basic hours	Total overtime hours
Monday	15/3	8.00 am	½ hr	4.30 pm	8	
Tuesday	16/3	8.00 am	½ hr	6.30 pm	8	2
Wednesday	17/3	8.00 am	½ hr	6.00 pm	8	1 ½
Thursday	18/3	8.00 am	½ hr	6.30 pm	8	2
Friday	19/3	8.00 am	½ hr	4.30 pm	8	
Saturday						
Sunday						
Signature: _JAMES_					40	5 ½

Figure 5.37

Employee's time sheet

Overtime is not normally paid to salaried persons. However they may receive an annual bonus or other performance-related payment. This is often based on the annual profits made by their employer.

Commission is paid to sales people. They might receive a basic wage or salary, plus a percentage of the value of the goods sold.

Deductions are made from wages or salaries for income tax, national insurance and pension fund payments etc.

◆ The wage or salary before deductions is known as the **gross amount**.
◆ The wage or salary after deductions is known as the **net amount** or 'take-home pay'.

```
PAY-ADVICE      : BSS.plc.SALARIED.PAYROLL          TAX PERIOD      : 07
PAY-DATE        : 26/10/2002                        N.I. NUMBER     : DB/06/78/01/B
EMPLOYEE REF    : NOTT/MANF/NADM/005695             N.I. TABLE      : F
EMPLOYEE NAME   : A. C. WHITEMAN                     TAX CODE        : 453L
PAY METHOD      : B.A.C.S.                          BASE RATE       : 18706 P.A.
OCCUPATION      : JOINERY MANAGER
```

-----------ALLOWANCES-------------	------------DEDUCTIONS--------------	---------------YTD TOTALS---------------
BASIC SALARY GROSS 1558.83	PAYE TAX 234.19	PAYE TAX 1638.95
	NAT. INS. 99.19	NAT. INS. 694.33
	SOCIAL CLUB 0.86	TAXABLE PAY 10696.49
	EES NI REBATE −1.06	SAVE FSC 0.00
	DEFINED CONTRI 30.76	LOAN BALANCE B 0.00
		PENSION FSC 0.00
		AVC FSC 0.00
		DEFINED C FSC 215.32
TOTAL ALLOWANCES 1558.83	TOTAL DEDS 363.94	**NET PAY** 1,194.89

Figure 5.38

Employees salary statement (payslip)

Ratio and proportions

Ratios and proportions are ways of comparing or stating the relationship between two similar or related quantities.

A bricklaying mortar mix may be described as 1:6 ('one to six'). This means the ratio or proportion of the mix is 1 part cement to 6 parts sand (or aggregate).

Figure 5.39

1 : 6 ratio

You could use bags, buckets or barrows as the unit of measure, and providing the proportions are kept the same they will make a suitable mortar mix. All that changes is the volume of mortar produced.

To share a quantity in a given ratio:

◆ add up the total number of parts or shares
◆ work out what one part is worth
◆ work out what the other parts are worth.

> **example**
>
> If £72 is to be shared by two people in a ratio of 5:3, what will each receive?
>
> **Number of shares = 5 + 3 = 8**
>
> **One share = 72 ÷ 8 = £9**
>
> **Five shares = 5 × 9 = £45**
>
> **Three shares = 3 × 9 = £27**
>
> **Answer: They will receive £45 and £27 respectively.**

> **example**
>
> A 1:3 ('one in three') pitched roof has a span of 3.6 m: what is its rise? This means for every 3 m span the roof will rise 1 m.
>
> **Rise = Span ÷ 3**
>
> $$\begin{array}{r} 1.2 \\ 3\overline{)3.6} \end{array}$$
>
> **Answer: The rise is 1.2 m**

Percentages

This is a standard way of representing a portion or part of a total quantity. Percentage (%) means 'per hundred' or 'per cent'.

In the diagram below, what part or portion of the total do these four squares make? We can write this in four ways:

- ◆ **Proportion**: 4 in 80 of the squares have been filled in, or
- ◆ **Ratio**: ⁴/₈₀ of the squares have been filled in, or
- ◆ **Percentage**: 5% of the squares have been filled in, or
- ◆ **Decimal**: 0.05 of the squares have been filled in.

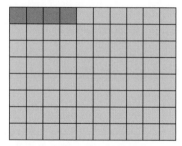

Figure 5.40

Percentage

Converting to a percentage

If you need to convert a number to a percentage:

- ◆ First divide the number by the total, i.e. ⁴/₈₀ = 0.05
- ◆ Then multiply by 100 to find the percentage, i.e. 0.05 × 100 = 5%.

Converting from a percentage

If you need to convert a percentage to a number:

- ◆ First divide the number by 100, i.e. 5% ÷ 100 = 0.05
- ◆ Then multiply by the total quantity, i.e. 0.05 × 80 = 4.

Hence 10% becomes 0.1; if the total number is 250, then 10% of 250 = 25.

Using percentages

There are three circumstances where percentages are used:

1. *Where a straightforward percentage of a number is required* – turn the percentage into decimal and multiply the total by it.

> **example**
>
> Find 12% of a total of 55 bags of cement.
>
> 55
> 0.12
> ─────
> 110
> 550
> ─────
> 660
>
> **ANSWER:**
> **6.6 bags**

2. **Where a number plus a certain percentage (increase) is required**
   – turn the percentage into decimal, place a one in front of it to include the original quantity and multiply by it.

> **example**
>
> Find 55 plus 12%
>
> 55 + 12% = 55 + 6.6 (see above example) = 61.6 bags.
>
> Alternatively, multiply 55 by 1.12:
>
> 55
> 1.12
> ─────
> 110
> 550
> 5500
> ─────
> 6160
>
> **ANSWER:**
> **61.6 bags**

3. **Where a number minus a certain percentage (decrease) is required**
   – take away percentage from 100, convert to decimal and then multiply by it.

> **example**
>
> Find 55 minus 12%
>
> 55
> 0.88
> ─────
> 440
> 4400
> ─────
> 4840
>
> **ANSWER: 48.4**

Statistics

We can make collections of numbers (statistics) speak to us in words and pictures. The presentation of statistics helps us to make sense of groups of numbers.

Averages

An average is the **mean value** of several numbers or quantities. It is found by adding the quantities and dividing by the number of quantities.

Average of the numbers = Sum of numbers ÷ Number of numbers.

> **example**
>
> To find the average of these 6 numbers: 2, 4, 6, 7, 9, 15, add them together and divide by 6:
>
> **Average = (2 + 4 + 6 + 7 + 9 + 15) ÷ 6**
> **= 43 ÷ 6**
> **= 7$\frac{1}{6}$**
> **= 7.167**

> **example**
>
> Find the average mark obtained for a number of college assessments: 48%, 27%, 49%, 75%, 84%, 44%, 65%.
>
> **Average = $\dfrac{48 + 27 + 49 + 75 + 84 + 44 + 65}{7}$**
>
> **= 392 ÷ 7**
> **= 56**
>
> **ANSWER: Average mark is 56%**

Mean, median and mode

These are all types of averages used in statistics:

◆ **Mean** is the true average we mainly refer to. It equals the sum of values divided by the number of values in the range, group or set of numbers.
◆ **Median** is the middle value when the numbers are put in order of size.
◆ **Mode** is the value that occurs most frequently.

> **example**
>
> The number of people late for work over a ten-day period was: 4, 2, 0, 1, 2, 3, 3, 6, 4, 2.
>
> Mean = (4 + 2 + 0 + 1 + 2 + 3 + 3 + 6 + 4 + 2) ÷ 10
> = 27 ÷ 10
> = 2.7 people late per day
>
> Median
> ◆ First put in size order: 0, 1, 2, 2, 2, 3, 3, 4, 4, 6
> ◆ Then cross off numbers from either end to find the middle value(s)
> ◆ The median is the middle value or the mean of the two middle values 2 and 3
> ◆ Median = 2.5 people late per day.
>
> Mode = 2 people late per day, because it occurs the most times.

Range of numbers

When using **mode** the range should be stated, to show how much the information is spread. In the above example:

Range = highest value − lowest value
$$= 6 - 0$$
$$= 6$$

Graphs

Graphs and charts are used to give a pictorial representation of numerical data.

Line graphs

These are used to show the relationship between two or more quantities.

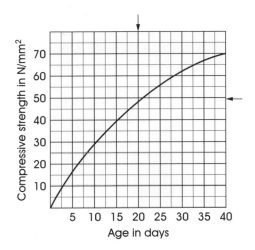

Figure 5.41

Line graph

It can be seen from the graph that concrete is expected to have a compressive strength of nearly 50 N/mm² after 20 days.

Bar charts

Bar charts use parallel bars (horizontal or vertical) to compare things or show change over time, the length of each bar being proportional to the quantity represented. Hence, by collecting data on work productivity, we could use our statistics to show how Friday has the lowest productivity of the week.

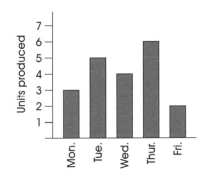

Figure 5.42

Bar charts

Grouped bar chart

This shows more than one set of data on the same chart and helps us to picture a large range of statistics. Hence we could compare the costs of wages and materials against profits over a number of years in the same chart.

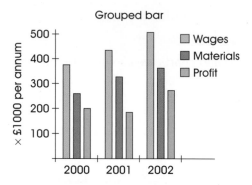

Figure 5.43

Grouped bar chart

Stacked bar chart

This shows the size of items in sub-groups or sub-sets of numbers, drawn in proportion to each other. Hence we could picture the number or percentage of different trades employed on-site at any one time.

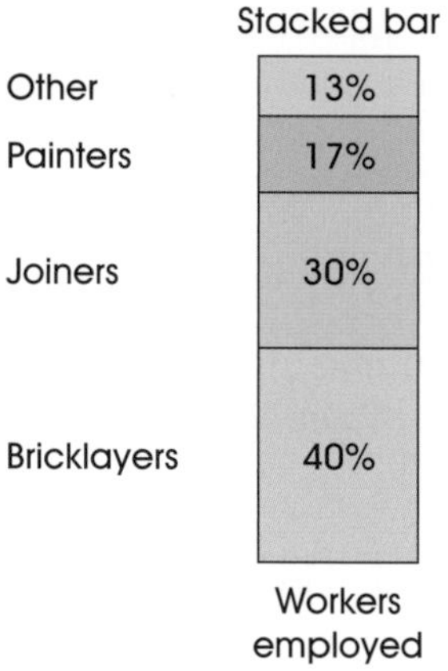

Stacked bar

Other	13%
Painters	17%
Joiners	30%
Bricklayers	40%

Workers employed

Figure 5.44

Stacked bar chart

Pie charts

These also show the proportional size of items that make up a set of data, pictured in the shape of a circle or 'pie'. Hence we could show pictorially how a company spends more of its earnings on wages than on its material costs.

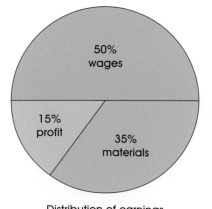

Distribution of earnings

Figure 5.45

Pie chart

Tally charts

These are used for collecting data. They show the frequency of occurrence of identical numbers or values in a range. A tally is just a group of lines or strokes, which are written down in fives for ease of counting. The fifth line forms a 'gate'.

Hence we could use a tally chart to show the number (frequency) and size (value) of critical measurements on factory-made components.

The following measurements were taken during a quality control check of 1.8-metre panel size lengths in a joinery shop. A tally chart was filled in by measuring the panel widths to five significant figures. The chart provides a quick observation of the accuracy of the manufacturing process.

1799.5	1799.5	1800	1800.5	1800.5
1800	1800	1800	1799	1799.5
1800.5	1800	1800	1799.5	1801
1799	1800	1800	1800.5	1800
1799.5	1799.5	1799.5	1800	1800
1800.5	1799	1800	1800	1800
1799.5	1800	1800	1800	1800.5
1799.5	1799.5	1799.5	1800	1800
1800	1799	1800	1800	1800.5
1799	1800	1800	1800.5	1799.5

Measurement	Tally	Frequency
1799	ЖЖ	5
1799.5	ЖЖ ЖЖ ||	12
1800	ЖЖ ЖЖ ЖЖ ЖЖ |||	23
1800.5	ЖЖ ||||	9
1801			1
Total	50	50

Histograms

A histogram or frequency diagram may be drawn to represent the frequency of occurrence determined in a tally chart. These look similar to bar charts except that there are no gaps between the bars.

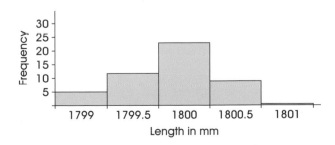

Figure 5.46

Histogram

Powers and roots of numbers

Powers

A simple way of writing repeated multiplications of the same number is shown below:

$$10 \times 10 = 100 \text{ or } 10^2$$
$$10 \times 10 \times 10 = 1000 \text{ or } 10^3$$
$$10 \times 10 \times 10 \times 10 = 10000 \text{ or } 10^4$$

and so on.

The small raised number is called the power or index. Numbers raised to the power 2 are usually called square numbers. We say 10^2 is '10 squared', and 10^3 is '10 cubed' but we say 10^4 is '10 to the power 4', etc.

Large numbers

Large numbers can therefore be written in a standard shorthand form by the use of an index or power:

$$30000000 = 3 \times 10000000 \text{ or } 3 \times 10^7$$
$$6600000 = 6.6 \times 1000000 \text{ or } 6.6 \times 10^6$$
$$990 = 9.9 \times 100 \text{ or } 9.9 \times 10^2$$

Small numbers

This standard form can also be used for numbers less than one, by employing a negative power or index:

$$0.036 = 3.6 \times 10^{-2} \text{ or } 36 \times 10^{-3}$$
$$0.0099 = 9.9 \times 10^{-3} \text{ or } 99 \times 10^{-4}$$
$$0.00012 = 1.2 \times 10^{-4} \text{ or } 12 \times 10^{-5}$$

The negative index is the number of places that the decimal point will have to be moved to the left if the number is written in full.

> **example**
>
> $99.0 \times 10^{-4} = 0.0099$

did you know?

The index number is the number of places that the decimal point has to be moved to the right if the number is written in full.

Roots

It is sometimes necessary to find a particular root of a number. Finding a root of a number is the opposite process of finding the power of a number. Hence, the **square root** is a number multiplied by itself once to give the number in question:

◆ The square of 5 is 5^2 or $5 \times 5 = 25$.
◆ Therefore the square root of 25 is 5.
◆ The common way of writing this is to use the square root sign $\sqrt{\ }$, e.g. $\sqrt{25} = 5$.

The **cube root** is a number multiplied by itself twice to give the number in question:

- The cube of 5 is 5^3 or $5 \times 5 \times 5 = 125$
- Therefore, the cube root of 125 is 5
- The common way of writing this is to use the root sign and an index of 3, i.e. $\sqrt[3]{}$, e.g. $\sqrt[3]{125} = 5$

From this we can see why there is a connection between powers and roots as opposite processes.

$$10^2 = 100 \qquad \sqrt{100} = 10$$
$$10^3 = 1000 \qquad \sqrt[3]{1000} = 10$$

Estimating roots

Roots can be found by estimation.

For example, $\sqrt{58.2}$ lies between the whole number squares of 49 (7×7) and 64 (8×8). So the $\sqrt{58.2}$ will be a decimal number between 7 and 8. And 64 is closer to 58.2 than 49, so the root will be closer to 8 than 7.

- Try $7.6 \times 7.6 = 57.76$; this is close but too small
- Try $7.7 \times 7.7 = 59.29$; this is close but too big.

57.76 is closer to 58.2 than 59.29 so the root will be closer to 7.6 than 7.7.

- Try $7.62 \times 7.62 = 58.0644$; this is close but too small
- Try $7.63 \times 7.63 = 58.2169$; this is close but slightly too big.

However 7.63 is a closer estimate than 7.62 and also correct to 1 decimal place.

Thus $\sqrt{58.2} = 7.63$ to 1 decimal place.

Calculating roots

Roots are more easily found using a calculator with a $\sqrt{}$ function key.

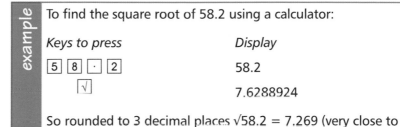

example

To find the square root of 58.2 using a calculator:

Keys to press	Display
5 8 · 2	58.2
√	7.6288924

So rounded to 3 decimal places $\sqrt{58.2} = 7.269$ (very close to our estimate).

Angles and lines

Angles

An angle is the amount of space between two intersecting straight lines. The angle does not depend on the length of the lines or the distance from the intersection that is measured. The size of an angle is measured in degrees (°).

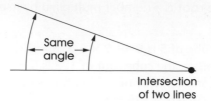

Figure 5.47

An anlge is the space between two interesection lines

Angles are also a measure of turning. For example, standing on a spot looking in one direction, if you were to turn completely round to look again in the same direction, you would have turned through 360° (degrees).

◆ A three-quarter turn is 270°
◆ A half turn is 180°
◆ A quarter turn is 90°. This is called a right angle.

Simple names and angle properties

◆ A **right angle** is 90°. They are shown on drawings by either a small square in the corners, or an arc with arrow heads and the figure 90°.

Figure 5.48

Angles are also a measure of turning

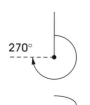

Figure 5.49

A right angle is 90°

◆ An **acute angle** is between 0° and 90°.

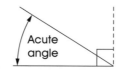

Figure 5.50

An acute angle is between 0° and 90°

◆ An **obtuse angle** is between 90° and 180°.

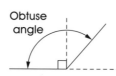

Figure 5.51

An obtuse angle is between 90° and 180°

◆ A **reflex angle** is between 180° and 360°.

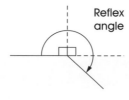

Figure 5.52

A reflex angle is between 180° and 360°

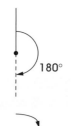

The addition of angles

Angles on a **straight line** always add up to 180°: A + B + C = 180°.

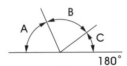

Figure 5.53

Straight line

Angles in a **triangle** (three-sided figure) always add up to 180°:
A + B + C = 180°.

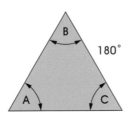

Figure 5.54

Triangle

Angles at a **point** always add up to 360°: A + B + C = 360°.

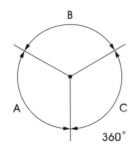

Figure 5.55

Point

Angles in a **quadrilateral** (four-sided figure) always add up to 360°:
A + B + C + D = 360°.

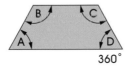

Figure 5.56

Quadrilateral

Angles in a **polygon** (more than four sides) always add up to 360°:
A + B + C + D + E = 360°.

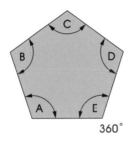

Figure 5.57

Polygon

Vertically opposite angles are always equal:

A = B; C = D

A + C = B + D = 180°

A + C + B + D = 360°

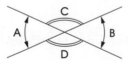

Figure 5.58

Lines and angles

Parallel lines are always the same distance apart: they never meet.

Figure 5.59

They are shown on a drawing using either a single or double pair of arrow heads.

Figure 5.60

A line drawn across a pair of parallel lines is called a **traverse**.

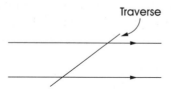

Figure 5.61

Alternate angles are always equal.

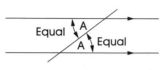

Figure 5.62

A = A

Corresponding angles are always equal.

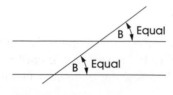

Figure 5.63

B = B

Supplementary angles always add up to 180°.

C + D = 180°

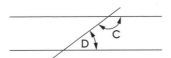

Figure 5.64

Measuring the size of angles

Reading angles

If asked to find the angle between AB and CB (called angle ABC), it means you have to find the angle of the middle letter, B.

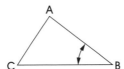

Figure 5.65

Protractors

Measuring angles can be done with the aid of a protractor.

Figure 5.66

◆ Place the protractor with its centre point over the intersection of the lines AB and CB and its baseline over one of the lines.
◆ Read off the angle at the edge of the protractor. It is 40° or 140° and is obviously less than 90°, so it must be 40° in this case.

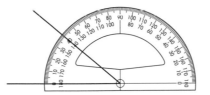

Figure 5.67

Shapes and solids

Shapes

A plane

A plane is a flat surface. Its shape is formed by lines known as the sides of the shape giving the length and breadth. The total distance all the way round the sides is called the perimeter. Plane shapes are classified by their number of sides.

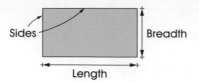

Figure 5.68
Properties of a plane figure

Triangle

A triangle is a plane shape that is bounded by three straight lines:

◆ The vertex is the angle opposite the base
◆ The altitude is the vertical height from the base to the vertex.

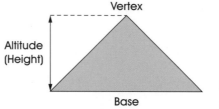

Figure 5.69
Properties of a triangle

Triangles are classified by either the length of their sides or by the size of their angles:

◆ An **equilateral triangle** has three equal length sides and three equal angles
◆ An **isosceles triangle** has two sides of equal length
◆ A **scalene triangle** has sides that are all unequal in length
◆ A **right-angled triangle** has one 90° angle
◆ In an **acute-angled triangle** all angles are less than 90°
◆ In an **obtuse-angled triangle** one of the angles is between 90° and 180°.

Quadrilateral

A quadrilateral is a plane shape that is bounded by four straight lines. A straight line joining opposite angles is called a diagonal and divides the figure into two triangles:

◆ A **square** has all four sides of equal length and all angles are right angles
◆ A **rectangle** has opposite sides of equal length and four right angles
◆ A **rhombus** has four equal sides, opposite sides being parallel, but none of the angles is a right angle
◆ A **parallelogram** has opposite sides that are parallel and equal in length, but one of its angles is a right angle
◆ A **trapezium** has two parallel sides
◆ A **trapezoid** has no parallel sides.

Equilateral triangle

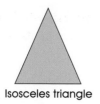

Isosceles triangle

Scalene triangle

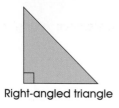

Right-angled triangle

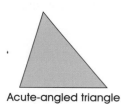

Acute-angled triangle

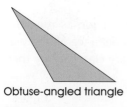

Obtuse-angled triangle

Figure 5.70 *Triangles*

Diagonal

Square

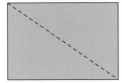

Rectangle

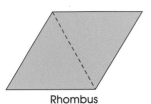

Rhombus

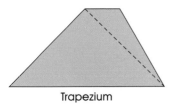

Parallelogram

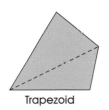

Trapezium

Trapezoid

Figure 5.71

Quadrilaterals

Polygon

A polygon is a plane shape that is bounded by more than four straight lines. Polygons may be classified as either:

◆ *Regular polygon* – these have sides of the same length and equal angles
◆ *Irregular polygon* – these have sides of differing length and unequal angles.

Both regular and irregular polygons are further classified by the number of sides they consist of:

◆ A **pentagon** consists of five sides
◆ A **hexagon** consists of six sides
◆ A **heptagon** consists of seven sides
◆ An **octagon** consists of eight sides.

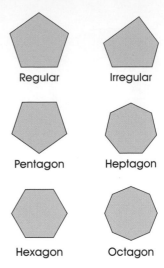

Regular   Irregular

Pentagon   Heptagon

Hexagon   Octagon

Figure 5.72 *Polygons*

Circles

A circle is a plane shape, bounded by a continuous, curved line, which at every point is an equal distance from the centre. The main elements of a circle are as follows:

◆ *Circumference* – the curved outer line (perimeter) of the circle.
◆ *Diameter* – a straight line that passes through the centre and is terminated at both ends by the circumference.
◆ *Radius* – the distance from the centre to the circumference. The radius is always half the length of the diameter.
◆ *Chord* – a straight line that touches the circumference at two points but does not pass through the centre.
◆ *Arc* – any section of the circumference
◆ *Normal* – any straight line that starts at the centre and extends beyond the circumference.
◆ *Tangent* – a straight line that touches the circumference at right angles to the normal.
◆ *Sector* – the portion of a circle contained between two radii and an arc. (Radii is the plural of radius.)
◆ *Quadrant* – a sector whose area is equal to a quarter of the circle.
◆ *Segment* – the portion of a circle contained between an arc and a chord.

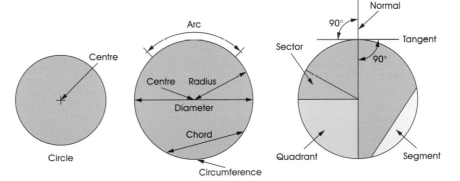

Figure 5.73

Properties of a circle

Figure 5.74

Annulus

Other circular features are:

◆ *Annulus* – the area of a plane shape that is bounded by two circles, each sharing the same centre but having different radii.
◆ *Ellipse* – a plane shape bounded by a continuous curved line drawn round two points called foci. The longest diameter is known as the major axis and the shortest diameter is the minor axis.

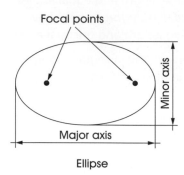

Ellipse

Figure 5.75

Properties of an ellipse

Solids

Solid figures are three-dimensional; they have length, breadth and thickness.

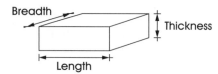

Figure 5.76

Properties of a solid

Prism

A solid figure contained by plane surfaces, which are parallel to each other. If cut into slices, they would all be the same shape.

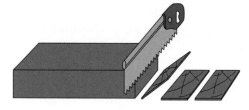

Figure 5.77

Slices of a prism are the same shape

All prisms are named according to the shape of their ends:

◆ *Cube* – all sides are equal in length and each face is a square.
◆ *Cuboid or rectangular prism* – each face is a rectangle and opposite faces are the same size. Bricks and boxes are examples of cuboids.
◆ *Triangular prism* – the ends are triangles and other faces are rectangles.
◆ *Octagonal prism* – the ends are octagons and other faces are rectangles.

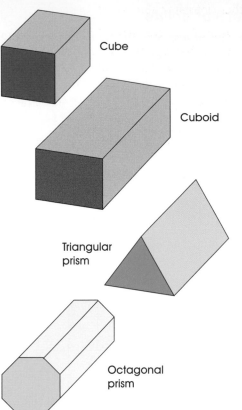

Figure 5.78
Prisms

Cube

Cuboid

Triangular prism

Octagonal prism

Apex

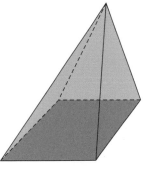

Triangular pyramid

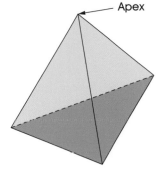

Square pyramid

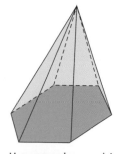

Hexagonal pyramid

Figure 5.79
Pyramids

Pyramid

A solid figure contained by a base and triangular sloping sides. The sides meet at a point called the apex. All pyramids are named according to their base shape:

- ◆ Triangular pyramid
- ◆ Square pyramid
- ◆ Hexagonal pyramid.

Circular solids

- ◆ *Cylinder* – a circular prism, the ends are circular in shape. Most tins are cylindrical.
- ◆ *Cone* – a circular pyramid, the base is circular in shape.
- ◆ *Sphere* – a solid figure, where all sections are circular in shape. Most balls are spherical.

Figure 5.80 *Circular solids*

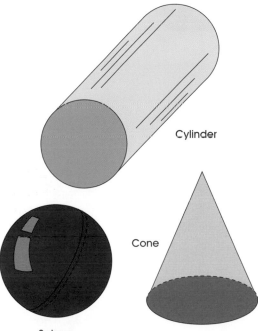

Cylinder

Cone

Sphere

Nets of solids

The net of a solid is the two-dimensional (2-D) shape, which can be folded to make the three-dimensional (3-D) shape.

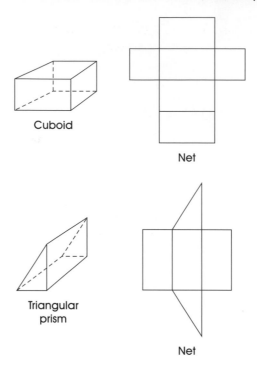

Cuboid

Net

Triangular prism

Net

Figure 5.81

Nets

measuring up

25. Convert 3.45 PM to the 24-hour clock.

26. Jane gets paid £7.55 per hour.
    a) How much will she receive for a 40-hour week?
    b) How much is this per annum?

27. A concrete mix is expressed as 1:3:6, being the proportion of cement, fine aggregate and course aggregate. How much of each is required for 2.4 m³?

28. Jason earns 75% of Julie's salary. How much does Jason earn if Julie earns £18 250 per annum?

29. Find the mean, median, mode and range of the following set of data: 8, 7, 8, 6.5, 10, 8, 7, 7.5, 9, 9, 8, 7.5, 9, 10, 8, 8.5, 9.5, 12.

30. How many degrees are there in two right angles?

31. How many degrees are there in a circle?

32. Angle ABC is 35 degrees. Which of the three angles does this refer to?

33. How many degrees does each angle measure in an equilateral triangle?

34. What is the name of a prism where all sides are equal in length and each face is a square?

35. What shape is a brick?

36. The pie chart shows the proportion of costs on a building job. Find the cost of the brickwork, if the total cost was £8762.50.

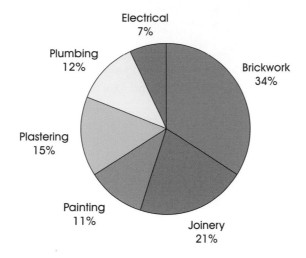

Figure 5.82

Numerical skills   Chapter 5

Formulae

Formulae are normally stated in algebraic terms. Algebra uses letters and symbols instead of numbers to simplify statements and enable general rules and relationships to be worked out.

Transposition of formulae

When solving a problem, sometimes formulae have to be rearranged in order to change the subject of the formulae before the calculation is carried out. Basically anything can be moved from one side of the equals sign to the other by changing its symbol. This means that on crossing the equals sign:

- Plus changes to minus
- Multiplication changes to division
- Powers change to roots.

This is also true vice versa.

Alternatively, we can cross-multiply. This means that on crossing the equals sign anything on the top line moves to the bottom line, and conversely anything on the bottom line moves to the top line.

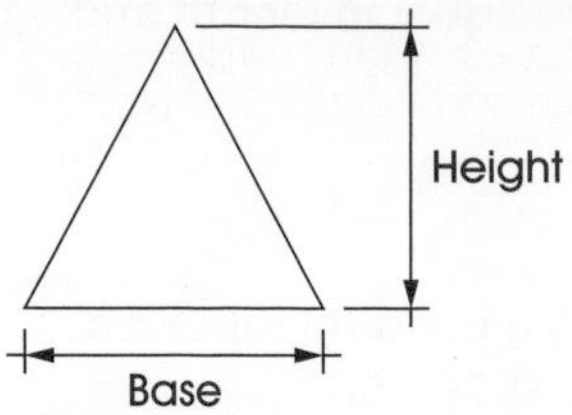

Figure 5.83

Area of triangle

Triangles

Area

The area of a triangle can be found by using the formula:
Area = Base × Height ÷ 2.

If we say:

- ◆ Area = A
- ◆ Base = B
- ◆ Height = H.

Then: A = B × H ÷ 2. Using algebra, we can abbreviate by missing out the multiplication sign (×) and expressing a division in its fractional form.

$$A = \frac{BH}{2} \text{ means the same as}$$

$$A = B \times H \div 2$$

---

example

Find the area of a triangle having a base of 2.5 m and a height of 4.8 m.

$$\text{Area} = \text{ Base } \times \text{ Height } \div 2$$

$$A = \frac{B \times H}{2}$$

$$= \frac{2.5 \times 4.8}{2}$$

$$= \frac{12}{2}$$

$$= 6$$

ANSWER: Area = 6 m²

Height

If the area and base of a triangle are known but we wanted to find out its height, the formula could be transposed to make the height the subject.

$$\text{Area} = \text{Base} \times \text{Height} \div 2$$

$$A = \frac{BH}{2}$$

Then $\frac{A}{B} = \frac{H}{2}$

(B moves from above to below on crossing the = sign)

And $\frac{2A}{B} = H$

(2 moves from below to above on crossing the = sign)

Therefore, Height = 2 × Area ÷ Base

> *example*
>
> Find the height of a triangle having an area of 4.5 m and a base of 1.5 m.
>
> $$\text{Area} = \frac{\text{Base} \times \text{Height}}{2}$$
>
> $$A = \frac{B\,H}{2}$$
>
> $$\frac{2A}{B} = H$$
>
> $$\frac{2 \times 4.5}{1.5} = H$$
>
> $$6 = H$$
>
> **ANSWER: Height = 6 m**

Rectangles

Perimeter

The perimeter of a rectangle can be found by using the formulae:

Perimeter = 2 × (Length + Breadth)

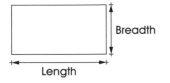

Figure 5.84

Perimeter of rectangle

> *example*
>
> Find the perimeter of a rectangle having a length of 3.6 m and a breadth of 2.2 m.
>
> $$\begin{aligned}\textbf{Perimeter} &= \textbf{2} \times \textbf{(Length + Breadth)} \\ &= \textbf{2 (L + B)} \\ &= \textbf{2 (3.6 + 2.2)} \\ &= \textbf{2 (5.8)} \\ &= \textbf{11.6}\end{aligned}$$
>
> **ANSWER: Perimeter = 11.6 m**

This can be abbreviated to: P = 2(L + B).

Plus and minus (+ and −) signs cannot be abbreviated and must always be shown in the formulae. For the correct order of working you must use the 'BODMAS' rule (see page 188). To obtain the correct answer, L must be added to B before multiplying by 2.

Circles

Perimeter and diameter

The formula for the perimeter or circumference of a circle is:
Circumference = π × Diameter.

C = π × D

π (spoken as pi) is the number of times that the diameter will divide into the circumference. It is the same for any circle and is taken to be 3.142.

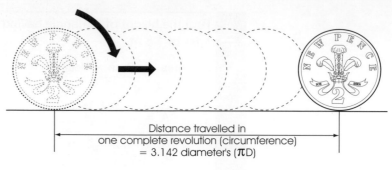

Distance travelled in
one complete revolution (circumference)
= 3.142 diameter's (πD)

Figure 5.85

*Relationship between diameter
and circumference*

Suppose we were given the circumference and asked to find the diameter.
Since C = π × D, then C ÷ π = D π is moved across = and changes from
× to ÷). Hence:

Diameter = Circumference ÷ π

> *example*
>
> Find the diameter of a circle having a circumference of 7.855 m.
>
> **Circumference** $=$ π × **Diameter**
>
> $$C = \pi \times D$$
>
> $$C \div \pi = D$$
>
> $$= 7.855 \div 3.142$$
>
> $$= 2.5$$
>
> **ANSWER:** <u>Diameter = 2.5 m</u>

Formulae for areas and perimeters

Values of **common shapes** can be found by using the following formulae:

- The perimeter of a figure is the distance or length around its boundary,
  linear measurement, given in metres run m.
- The area of a figure is the extent of its surface, **square measurement**,
  given in square metres (m²).

Table 5.2

Shape	Area equals	Perimeter equals
Square	AA	4A
Rectangle	LB	2(L + B)
Parallelogram	Area equals BH	Perimeter equals 2(A + B)

Table 5.2 (cont.)

Shape		Area equals	Perimeter equals
Trapezium	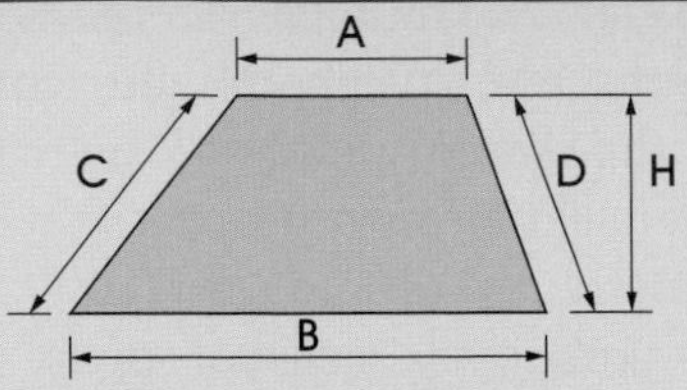	$\dfrac{(A + B)H}{2}$	$A + B + C + D$
Triangle	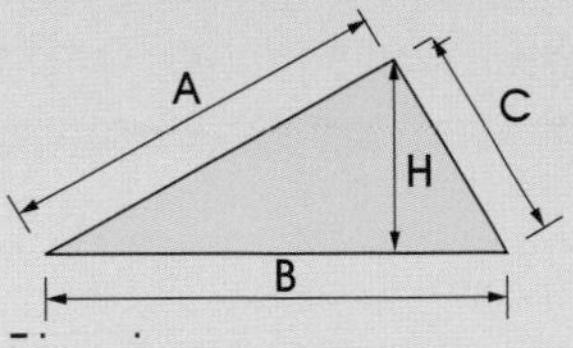	$\dfrac{BH}{2}$	$A + B + C$
Circle	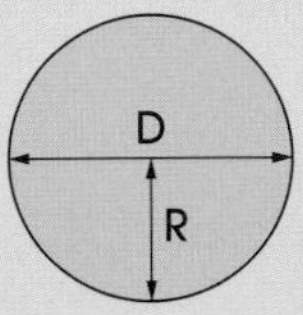	πR^2	πD or $2\pi R$
Annulus	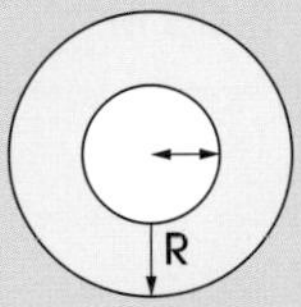	$\pi R^2 - \pi r^2$	$2\pi R + 2\pi r$
Sector	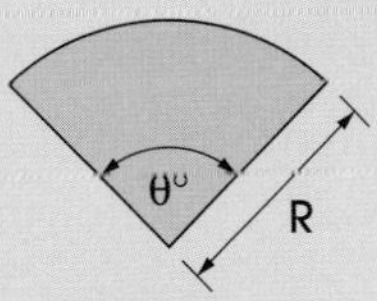	$\dfrac{\theta^\circ}{360}\pi R^2$	$\dfrac{\theta^\circ}{360}2\pi R$ (Arc only)
Ellipse	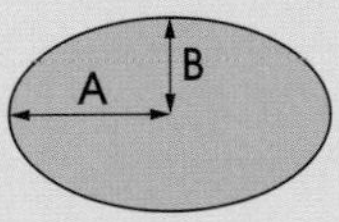	πAB	$\pi(A + B)$

Complex areas

Complex areas can be calculated by breaking them into a number of recognizable areas and solving each one in turn, e.g. the area of the room shown in the figure is equal to area A plus area B minus area C.

example

Find the area of the room above

Area A $= \left(\dfrac{9 + 10.5}{2}\right) \times 9.75$

$= 95.813 \text{ m}^2$

Area B $= 0.75 \times 5.5$

$= 4.125 \text{ m}^2$

Area C $= 0.9 \times 3$

$= 2.7 \text{ m}^2$

Total area $\quad A + B - C$

$95.813 + 4.125 - 2.7$

97.238 m^2

ANSWER: Area = <u>97.238 m²</u>

Converting to the same units

We can only multiply like terms. Where metres and millimetres are contained in the same problem, first convert the millimetres into a decimal part of a metre by dividing by 1000. (Move the imaginary decimal point behind the number three places forward.) Alternatively, convert all units to millimetres.

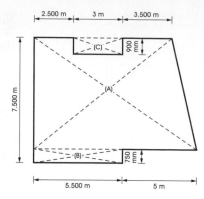

Figure 5.86 *Complex area*

example

Convert 50 mm to metres.

.050
3 2 1

ANSWER:

50 mm = 0.05 m

Volume

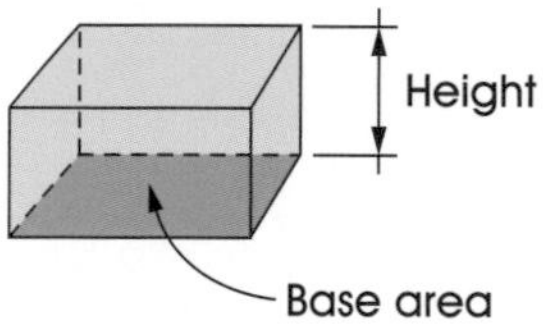

The volume of an object can be defined as the space it takes up, **cubic measurement**, given in cubic metres (m³).

Many solids have a uniform cross-section and parallel edges. The volume of these can be found by multiplying their base area by their height:

Volume = Base area × Height

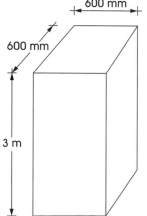

Figure 5.87 *Volume*

example

Find the volume of concrete required for a 600 mm square, 3-metre high column

Volume = Base area × Height
= 0.6 × 0.6 × 3
= 1.08 m³

ANSWER: Volume = 1.08 m³

example

A house contains forty-eight 50 mm × 225 m softwood joists, 4.5 m long. How many cubic metres of timber are required?

Volume of 1 joist = 0.05 × 0.225 × 4.5
= 0.050625 m³
Total volume = 0.050625 × 48
= 2.43 m³

ANSWER: 2.43 m³ of timber are required

Figure 5.88 *Concrete column*

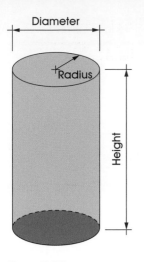

Diameter

Radius

Height

Figure 5.89
Surface area

Surface area

The lateral surface area of a solid with a uniform cross-section is found by multiplying its base perimeter by its height.

Lateral surface area = Base perimeter × Height

For a **cylinder**, the lateral surface area = π × Diameter × Height.

The **total surface area** can be determined by adding the areas of ends or base to the lateral surface area.

The total surface area of a cylinder = π × Diameter × Height + (π × Radius × Radius × 2), i.e. (Lateral area) × (Area of ends):

Surface area = $\pi DH + (2\pi R^2)$

---

example

Find the total surface area of a cylinder 1.2 metre diameter and 2.4 metres in height.

Total surface area

$$= \pi D \times H + (\pi R^2 \times 2)$$
$$= (3.142 \times 1.2 \times 2.4) + (3.142 \times 0.6 \times 0.6 \times 2)$$
$$= 9.04896 + 2.26224$$
$$= 11.3112 \ m^2$$

say 11.31m² to 2 decimal places

ANSWER: Total surface area = 11.31 m²

Formulae for volumes and surface areas

The following formulae can be used for calculating the volume and lateral surface area of frequently used **common solids**. It can be seen that the volume of any pyramid or cone will always be equal to one-third of its equivalent prism or cylinder.

Table 5.3

Shape	Volume	Lateral surface area
Rectangular prism	LBH	2(L+B)H
Rectangle pyramid	$\dfrac{LBH}{3}$	S(L + B)

Shape	Volume	Lateral surface area
Cylinder 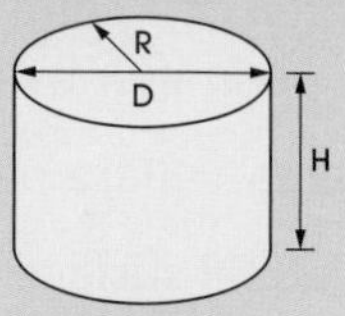	$\pi R^2 H$	$\pi D H$
Cone 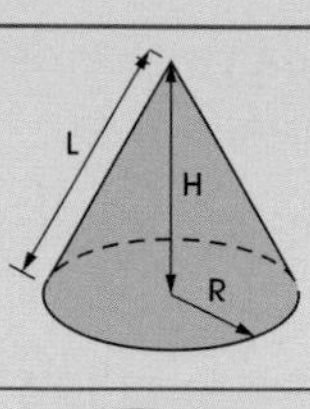	$\dfrac{\pi R^2 H}{3}$	$\pi R L$
Sphere 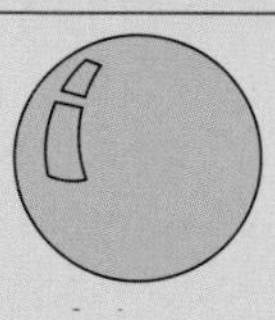	$\dfrac{4\pi R^3}{3}$	$4\pi R^2$

Complex volumes

These are found by breaking them up into a number of recognizable volumes and solving for each one in turn. This is the same as the method used when solving for complex areas.

Suppose you were asked to find the volume of concrete required for a 2.4-metre high column having the plan or horizontal cross-section shown in Figure 5.90.

The column can be considered as a rectangular prism (A) and half a cylinder (B).

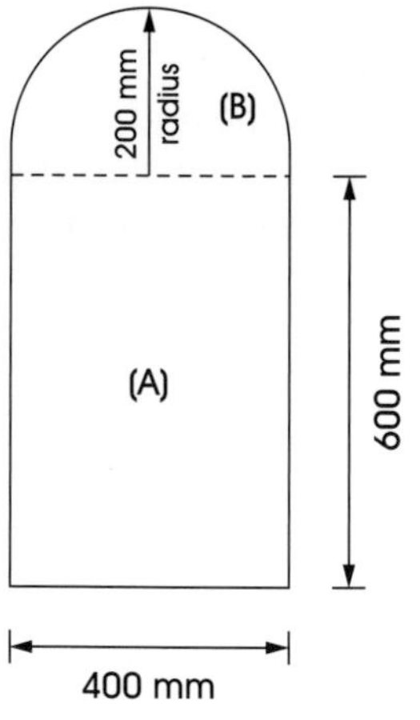

example

Find the volume of the column shown.

Volume A $= 0.4 \times 0.6 \times 2.4$

$= 0.576$ m³

Volume B $= \dfrac{3.142 \times 0.2 \times 0.2 \times 2.4}{2}$

$= 0.151$ m³

Total vol. $= 0.576 + 0.151$

$= 0.727$ m³

ANSWER: 0.727 m³ of concrete is required

Figure 5.90

Plan shape of column

Where a solid tapers, its volume can be found by multiplying its average cross-section by its height.

Volume = Average cross-section × Height

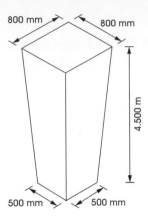

Figure 5.91

Tapered column

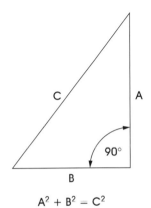

$$A^2 + B^2 = C^2$$

Figure 5.92

Pythagoras' theorem

example

Find the volume of the tapered column shown in Figure 5.91.

$$\text{Volume} = \frac{(0.8 \times 0.8) + (0.5 \times 0.5)}{2} \times 4.5$$

$$= \frac{0.64 + 0.25}{2} \times 4.5$$

$$= 2.003 \text{ m}^3$$

ANSWER: Volume = 2.003 m³

Pythagoras' theorem

The lengths of the sides in a right-angled triangle can be found using Pythagoras' theorem. According to this theorem, in any right-angled triangle the square of the length of the longest side is equal to the sum of the squares of the other two sides.

$$C^2 = A^2 + B^2$$

A simple version of Pythagoras' theorem is known as the **3:4:5 rule**. This is often used for setting out and checking right angles, since a triangle whose sides equal 3 units, 4 units and 5 units must be a right-angled triangle because $5^2 = 3^2 + 4^2$.

If we know the lengths of two sides of a right-angled triangle we can use Pythagoras' theorem to find the length of the third side. In fact this theorem forms the basis of pitched roof calculations.

example

Calculate the length of the common rafter shown in Figure 5.93.

A = Rise 2.1 m

B = Run 2.8 m

C = Common rafter

$$= A^2 + B^2 = C^2$$

$$2.1^2 + 2.8^2 = C^2$$

$$= 4.41 + 7.84 = 12.25$$

Therefore:

$$C = \sqrt{12.25} = 3.5 \text{ m}$$

ANSWER:

The common rafter is 3.5 m long

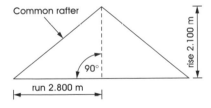

Figure 5.93

Roof section

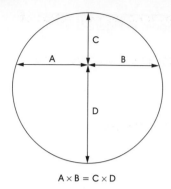

$$A \times B = C \times D$$

Figure 5.94

Interesecting chords rule

Intersecting chords rule

Where two chords intersect in a circle, the product (result of multiplication) of the two parts of one chord will always be equal to the product of the two parts of the other chord. It is very useful for finding radius lengths.

example

Suppose you were asked to set out a turning piece having a span of 1.2 metre and a rise of 100 metres. What is the radius?

A = 0.6 m

B = 0.6 m

C = 0.1 m

D = ?

$$A \times B = C \times D$$

$$\frac{A \times B}{C} = D$$

$$\frac{0.6 \times 0.6}{0.1} = 3.6$$

$$\text{Radius} = \frac{C + D}{2}$$

$$= \frac{0.1 + 3.6}{2}$$

$$= 1.85 \text{ m}$$

ANSWER: The radius is 1.85 m

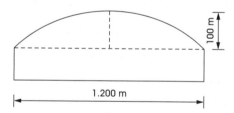

Figure 5.95

Turning piece

did you know?

A turning piece is a solid piece of timber cut to the shape of a flat or segmental arch, which is used to provide temporary support for the brickwork during construction.

Trigonometry

Trigonometry involves understanding the relationship between the sides and angles of right-angled triangles.

You will have to set up and solve equations to find unknown lengths or angles. There are three ratios to understand in right-angled triangles. These are sine, cosine and tangent ratios.

In trigonometry, the sides of a right-angled triangle are given temporary names in relation to the angle 'θ' being considered (see Figure 5.96).

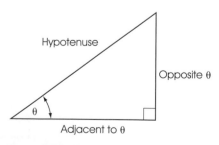

Figure 5.96 *Relationship between sides and angles used in trigonometry*

The ratios of these sides to each other are the **trigonometrical ratios** of the angle θ:

The **sine** of θ = Opposite side/Hypotenuse (longest side), or abbreviated:

$$\sin \theta = \frac{\text{opp}}{\text{hyp}}$$

The **cosine** of θ = Adjacent side/Hypotenuse, or abbreviated:

$$\cos \theta = \frac{\text{adj}}{\text{hyp}}$$

The **tangent** of θ = Opposite side/Adjacent side, or abbreviated:

$$\tan \theta = \frac{\text{opp}}{\text{adj}}$$

Using trigonometry to find the length of an unknown side

If we know two parts of the equation we can use trigonometry to find the unknown third part.

Your calculator may have different angle modes; you should set it to degrees. With DEG showing on the display enter 30 then the sin key. If the mode is correctly set it should display 0.5:

$$\sin 30° = 0.5$$

---

example 1

Find the length of side 'X'.

$$\text{Sin } 35° = \frac{X}{6}$$

$$6 \times \text{Sin } 35° = X$$

$$X = 3.441458618$$

$$= 3.441 \text{ m}$$

6 m   X m

35°

Figure 5.97

ANSWER:
The length of side **X** is **3.441 m**

---

example 2

Find the length of side 'X'.

$$\text{Tan } 40° = \frac{4.8}{X}$$

$$\frac{\text{Sin } 35°}{4.8} = X$$

$$X = 4.02767823$$

$$= 4.028 \text{ m}$$

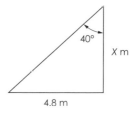

40°

X m

4.8 m

Figure 5.98

ANSWER:
The length of side **X** is **4.028 m**

In **Example 1** we know the hypotenuse and the angle opposite X. Thus use the sine ratio.

Transpose formula to get **X** on its own. Enter 35 sin × 6 = into the calculator and round answer to three decimal places.

In **Example 2** we know the angle adjacent to **X** and the length of the opposite side. Thus use the tangent ratio. Transpose formula to get **X** on its own. Enter 40 tan ÷ 4.8 = into the calculator and round to three decimal places.

Using trigonometry to finding an unknown angle

On your calculator you will find keys marked $\sin^{-1}$, $\cos^{-1}$ and $\tan^{-1}$. You may have to press an inverse INV key first. $\sin^{-1}$ means 'the angle whose sin is . . .'

$$\sin^{-1} 0.50 = 30°.$$

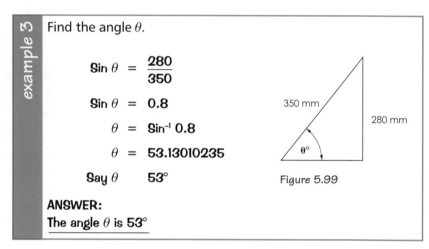

example 3

Find the angle θ.

$$\text{Sin } \theta = \frac{280}{350}$$

$$\text{Sin } \theta = 0.8$$

$$\theta = \text{Sin}^{-1} 0.8$$

$$\theta = 53.13010235$$

Say θ 53°

350 mm

280 mm

θ°

Figure 5.99

ANSWER:
The angle θ is **53°**

In **Example 3** we know the lengths of the opposite side and the hypotenuse. Thus use the sine ratio. Divide opposite by hypotenuse. Enter $0.8 \sin^{-1}$ = (or 0.8 INV sin =) into the calculator. Round the answer to nearest 0.5 of a degree.

measuring up

37. What is the relationship between the diameter and circumference of a circle?

38. A rectangle has a length of 6.39 m and a breadth of 2.15 m.
    a) What is the area?
    b) What is the perimeter?

39. What is the volume of a room if it measures 4.5 m × 3.65 m × 2.4 m?

40. What is the total surface area of a cylindrical prism if the radius is 600 mm and the height is 1.9 m?

41. The longest side in a right-angled triangle measures 5 m; one of the other sides is 3 m. What is the length of the remaining side?

42. Calculate the area and perimeter of the figure illustrated.

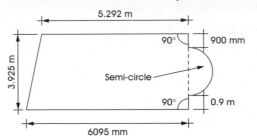

Figure 5.100

43. Find the length of the side marked X for the triangle illustrated.

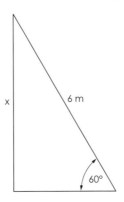

Figure 5.101

Measuring and costing materials

Flooring

In order to determine the amount of floor covering materials required for an area, multiply its width by its length.

Floorboards

137 mm covering width

Figure 5.102

Section of a floorboard

To calculate the metres run of floorboards required to cover a floor area of say 4.65 m², if the floorboards have a covering width of 137 mm:

$$\text{Metres run required} = \text{Area} \div \text{Width of board}$$
$$= 4.65 \div 0.137$$
$$= 33.94 \text{ m,}$$
$$\text{say } 34 \text{ m run}$$

It is standard practice to order an additional amount of flooring to allow for cutting and **wastage**. This is often between 10% and 15%.

If 34-metre run of floor boarding is required to cover an area, calculate the amount to be ordered including an additional 12% for cutting and wastage.

$$\text{Amount to be ordered} = 34 \times 1.12$$
$$= 38.08, \text{ say } 38 \text{ m run}$$

In order to determine the number of sheets of plywood or chipboard required to cover a room either:

◆ Divide area of room by area of sheet, or
◆ Divide width of room by width of sheet and divide length of room by length of sheet. Convert these numbers to the nearest whole or half and multiply them together.

To calculate the number of 600 mm × 2400 mm chipboard sheets required to cover a floor area of 2.02 m × 3.6 m.

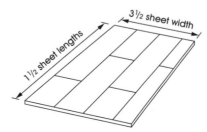

Figure 5.103

Area of floor

Number of sheets required	= Area of room ÷ Area of sheet
Area of room	= 2.05 m × 3.6 m
	= 7.38 m²
Area of sheet	= 0.6 m × 2.4 m
	= 1.44 m²
Number of sheets	= 7.38 × 1.44
	= 5.125, say 6 sheets

Alternatively:

Number of sheet widths in room width	= 2.05 ÷ 0.6
	= 3.417, say 3.5
Number of sheet lengths in room length	= 3.6 ÷ 2.4
	= 1.5
Total number of sheets	= 3.5 × 1.5
	= 5.25, say 6 sheets

Joists

To determine the number of joists required and their centres for a particular area the following procedure can be used.

◆ Measure the distance between adjacent walls, say 3150 mm.
◆ The first and last joist would be positioned 50 mm away from the walls. The centres of 50-mm breadth joists would be 75 mm away from the wall. The total distance between end joists centres would be 3000 mm. See Figure 5.104(a) below.
◆ Divide the distance between end joists centres by the specified joist spacing, say 400 mm. This gives the number of spaces between joists. Where a whole number is not achieved round up to the nearest whole number above. There will always be one more joist than the number of spaces, so add another one to this figure to determine the number of joists. See Figure 5.104(b) below.

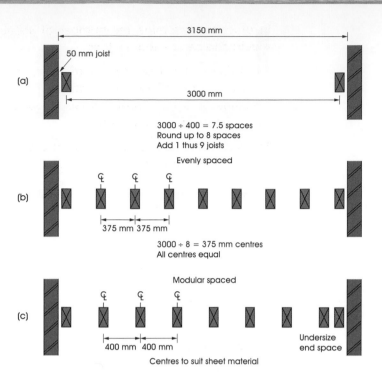

Figure 5.104

Determining number of joists

◆ Where T&G (tongue and groove) boarding is used as a floor covering the joist centres may be spaced out evenly, i.e. divide the distance between end joist centres by the number of spaces.

◆ Where sheet material is used as a joist covering to form a floor, ceiling or roof surface, the joist centres are normally maintained at a 400 mm or 600 mm module spacing to coincide with sheet sizes. This would leave an undersized spacing between the last two joists. See (c) above.

Number of joists
= (Distance between end joist centres x Joist spacing) + 1
= (3000 ÷ 400) + 1
= 8.5, say 9

Stud partitions

To determine the number of studs required for a particular partition the following procedure can be used.

◆ Measure the distance between the adjacent walls of the room or area, which the partition is to divide, say 3400 mm (see Figure 5.105).
◆ Divide the distance between the walls by the specified spacing, say 600 mm. This gives the number of spaces between the studs. Use the whole number above. There will always be one more stud than the number of spaces, so add one to this figure to determine the number of studs. Stud centres must be maintained to suit sheet material sizes leaving an undersized space between the last two studs.
◆ The lengths of head and sole plates are simply the distance between the two walls.
◆ Each line of noggins will require a length of timber equal to the distance between the walls.

Numerical skills

Chapter 5

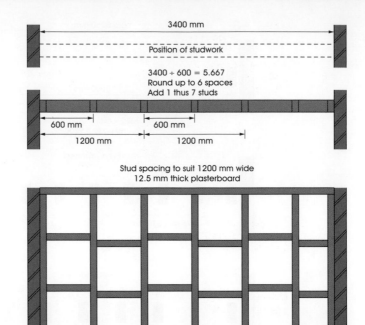

Figure 5.106

Determining number of studs

The total length of timber required for a partition can be determined by the following method:

7 studs at 2.4 mm; 7 × 2.4 = 16.8 m
Head and sole plates at 3.4 m; 2 × 3.4 = 6.8 m
3 lines of noggins at 3.4 m; 3 × 3.4 = 10.2 m
Total metres run required, 16.8 + 6.8 + 10.2 = 33.8 m, say 34 m

Rafters

The number of rafters required for a pitched roof can be determined using the same method as used for floor joists. For example, divide the distance between the end rafter centres by the rafter spacing. Round up and add one. Remember to double the number of rafters to allow for both sides of the roof.

Say distance between end rafter centres is 12 m and spacing is 400 mm:

Number of rafters
= Distance between end rafter centres ÷ Rafter spacing + 1
= 12 ÷ 0.4 + 1
= 30 + 1
= 31 rafters

Therefore total number of rafters required for both sides of the roof is 62.

Where overhanging verges are required an additional rafter must be allowed at each end to form gable ladders, which provide a fixing for the bargeboard and soffit.

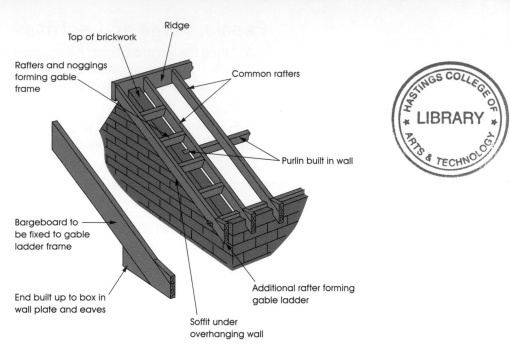

Figure 5.106

Pitch roof components

To determine the length of rafters Pythagoras' theorem of right-angled triangles can be used.

Determine the length of rafter required for a roof of 2.5 m rise and 6 m span.

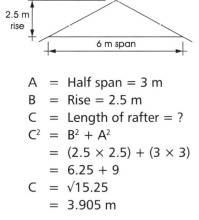

Figure 5.107

Pitched roof section

A = Half span = 3 m
B = Rise = 2.5 m
C = Length of rafter = ?
C^2 = $B^2 + A^2$
= $(2.5 \times 2.5) + (3 \times 3)$
= 6.25 + 9
C = $\sqrt{15.25}$
= 3.905 m

An allowance must be added to this length for the eaves overhang and the cutting. Say 0.5 m and 10%.

Total length of rafter = 3.905 + 0.5 + 10%
= 4.405 × 1.1
= 4.845 m

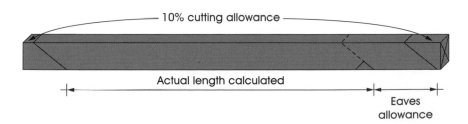

Figure 5.108

Rafter allowances

Numerical skills

Chapter 5

Fascia, barge and soffits

Calculating the lengths of material required for fascia boards, bargeboards and soffits is often a simple matter of measuring, allowing a certain amount extra for jointing, and adding lengths together to determine total metres run.

A hipped-end roof requires two 4.4-metre lengths and two 7.2-metre lengths of ex 25 mm × 150 mm PAR softwood for its fascia boards.

$$\text{Metres run required} = (4.4 \times 2) + (7.2 \times 2)$$
$$= 8.8 + 14.4$$
$$= 23 \text{ m}$$

It is standard practice to allow a certain amount extra for cutting and jointing. This is often 10%.

$$\text{Total metres run required} = 23.2 \times 1.1$$
$$= 25.52 \text{ m}$$

The length of timber for bargeboards may require calculation using Pythagoras' theorem for right-angled triangles.

Sheet material

Where sheet material is used for fascias and soffits the amount that can be cut from a full sheet often needs calculating. This entails dividing the width of the sheet by the width of the fascia or soffit, and then using the resulting whole number to multiply by the sheet's length, to give the total metres run.

example

Determine the total metres run of 150 mm wide soffit board that may be cut from a 1220 mm × 2440 mm sheet.

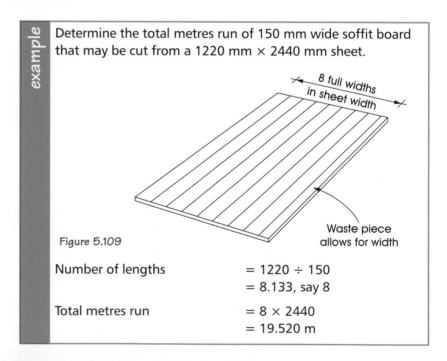

Figure 5.109

Number of lengths	$= 1220 \div 150$
	$= 8.133$, say 8
Total metres run	$= 8 \times 2440$
	$= 19.520 \text{ m}$

Trim

To determine the amount of trim required for any particular task is a fairly simple process, if the following procedures are used.

Architraves

The jambs or legs in most situations can be taken to be 2100 mm long. The head can be taken to be 1000 mm. These lengths assume a standard full-size door and include an allowance for mitring the ends. Thus the length of architrave required for one face of a door lining/frame is 5200 mm or 5.2 m.

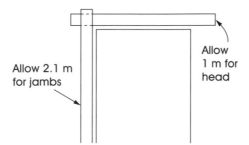

Allow 1 m for head

Allow 2.1 m for jambs

Figure 5.110 *Architrave*

Multiply this figure by the number of architrave sets to be fixed. This will determine the total metres run required, say 8 sets, both sides of four doors:

5.2 × 8 = 41.6 m.

Skirtings

Skirtings and other horizontal trim can be estimated from the perimeter. This is found by adding up the lengths of the walls in the area. The widths of any doorways and other openings are taken away to give the actual metres run required.

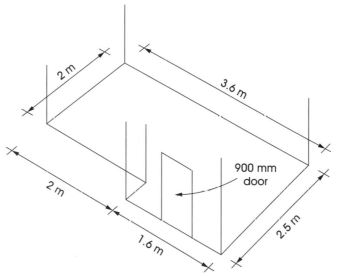

2 m

3.6 m

2 m

900 mm door

1.6 m

2.5 m

Figure 5.111 *Perimeter of room*

Determine the total length of timber required for the room.

Perimeter	= 2 + 3.6 + 2.5 + 1.6 + 0.5 + 2
	= 12.2 m

Total metres run required = 12.2 – 0.9 (door opening)
= 11.3 m

An allowance of 10% for cutting and waste is normally included in any estimate for horizontal moulding.

Determine the total metres run of skirting required for the run shown including an allowance of 10% for cutting and waste.

Total metres run required = 11.3 + 1.13
= 12.43 m, say 12.5 m

Brickwork and mortar

There are 60 bricks per m² in half brick thick walls and 120 bricks per m² in one brick thick walls. An additional percentage of 5% is normally allowed for cutting and damaged bricks.

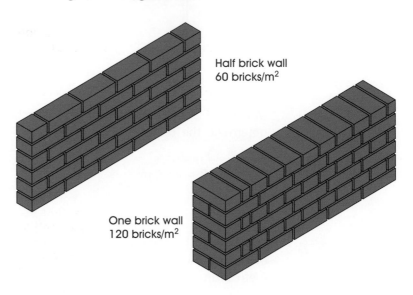

Half brick wall
60 bricks/m²

One brick wall
120 bricks/m²

Figure 5.112

Half and one brick walls

Approximately 1 kg of mortar is required to lay one brick. This figure can be used for small areas of new brickwork and making good. A 25 kg bag of mortar mix is sufficient to lay up to 25 bricks, depending on the thickness of joints. One 25 kg bag of cement and six 25 kg bags of sand is sufficient to lay up to 175 bricks using a 1:6 cement–sand ratio.

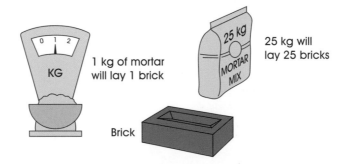

1 kg of mortar
will lay 1 brick

25 kg will
lay 25 bricks

Brick

Figure 5.114

Relationship between amount of mortar and number of bricks

For larger areas of brickwork the amount of mortar can be assessed as 0.03 m³ per square metre of brickwork or 0.5 m³ per 1000 bricks for half brick walls. Just over double the amount per square metre for one brick walls, 0.07 m³, or 0.58 m³ per 1000 bricks. An additional percentage of 10% is normally allowed for wastage.

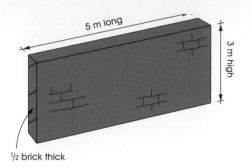

Figure 5.114

Area of half brick wall

¹/₂ brick thick

Calculate the number of bricks and amount of mortar required for a half brick wall 3 m high and 5 m long, allow 5% extra for cutting and 10% for mortar.

Area of wall	= Height × Length
	= 3 × 5
	= 15 m²
Number of bricks required	= Area × Number of bricks per m²
	= 15 × 60
	= 900 bricks
5% allowance	= 900 × 1.05
	= 945 bricks
Amount of mortar	= Area × Mortar/m²
	= 15 × 0.03
	= 0.45 m³
10% allowance	= 0.45 × 1.1
	= 0.495 m³, say 0.5 m³

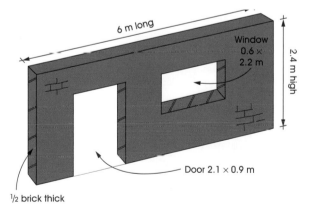

Figure 5.115

Area of half brick wall with openings

¹/₂ brick thick

Calculate the number of bricks and the amount of mortar required to form a half brick wall 2.4 m high and 6 m long containing a 2.1 m × 0.9 m door opening and a 0.6 m × 2.2 m window opening.

Area of brickwork = Total area − Area of door and window

Total area = 2.4 × 6
= 14.4 m²

Area of door = 2.1 × 0.9
= 1.89 m²

Area of window = 0.6 × 2.2
= 1.32 m²

Therefore:

Area of brickwork = 14.4 − 1.89 − 1.32
= 11.19 m²

No. of bricks = Area × No. of bricks/m²
= 11.19 × 60
= 671.4, say 672

5% allowance = 672 × 1.05
= 705.6, say 706

Amount of mortar = Area × Mortar/m²
= 11.19 × 0.03
= 0.336 m³

10% allowance = 0.336 × 1.1
= 0.369 m³

Where brickwork returns around corners as in a building the wall's centre line length is used to calculate the area.

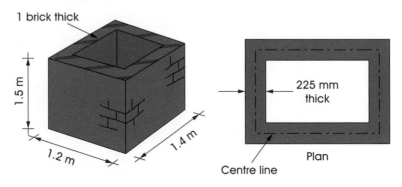

Figure 5.116

Brickwork inspection chamber

did you know?

An 'inspection chamber' provides access to underground drain intersections and sharp bends. These permit access for visual inspection and clearing of blockages. Inspection chambers are also called 'manholes'.

Calculate the number of bricks and amount of mortar to build the one-brick thick inspection chamber.

Length of centre line
= (2 × Length) + (2 × Width) − (4 × Wall thickness)
= 2.8 + 2.2 − 0.9
= 4.1 m

Area = Length of centre line x Height
= 4.1 × 1.5
= 6.15 m²

Number of bricks = Area × Number of bricks per m²
= 6.15 × 120
= 738

5% allowance   = 738 × 1.05
                    = 774.9, say 775 bricks

Amount of mortar = Area × Mortar per m²
                    = 6.15 × 0.07
                    = 0.4305 m³

10% allowance   = 0.4305 × 1.1
                    = 0.4735, say 0.474 m³

Blockwork and mortar

There are approximately ten 450 mm × 215 mm blocks per square metre. An additional allowance of 5% may be added for cutting and damaged blocks. Each m² of 100 mm thick blocks can be assessed as requiring 0.01 m³ of mortar. An allowance of 10% is normally added for wastage.

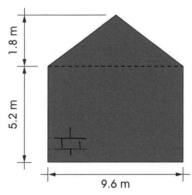

Figure 5.117

Gable end

Calculate the number of 450 mm × 215 mm blocks and mortar required for the gable end illustrated. Allow 5% extra on the blocks and 10% extra on the mortar.

Area of blockwork = Triangular area + Rectangular area

Area of triangle   = Base × Height ÷ 2
                    = 9.6 × 1.8 ÷ 2
                    = 8.64 m²

Area of rectangle = Length × Height
                    = 9.6 × 5.2
                    = 49.92 m²

Total area   = 8.6 + 49.92
                    = 58.56 m²

Number of blocks   = Area × Number of blocks per m²
                    = 58.56 × 10
                    = 585.6 blocks

5% allowance   = 585.6 × 1.05
                    = 614.8, say 615 blocks

Amount of mortar = Area × Mortar per m²
                    = 58.56 × 0.01
                    = 0.586 m³

10% allowance   = 0.586 × 1.1
                    = 0.644 m³

Blockwork walls with returns and openings are calculated in the same ways as those in brickwork using the centre line length when working out the area.

Dry materials

Mortar

The density of mortar is approximately 2300 kg/m³. Different mixes are specified for different situations.

A 1:6 mix contains 7 parts:

- ◆ 1 part cement
- ◆ 6 parts sand (fine aggregate).

A 1:1:8 mix contains 10 parts:

- ◆ 1 part cement
- ◆ 1 part lime
- ◆ 8 parts sand.

To determine dry materials for a given quantity of mortar, multiply volume by density 2300 kg/m³ for total mass and divide by number of parts. This gives the amount of cement; multiply cement by required number of parts to give other quantities.

For 0.5 m³ of 1:6 mortar:

Mass of mortar $= 0.5 \times 2300$
$= 1150$ kg

Number of parts in mix $= 7$

Amount of cement $= 1150 \div 7$
$= 164.286$ kg, say seven 25 kg bags

Amount of sand $= 164.286 \times 6$
$= 985.716$ kg, say 1 tonne or 40×25 kg bags

Mixed on-site concrete

It is necessary to determine the amount of cement, fine and coarse aggregates for ordering. The density of compacted concrete is 2400 kg/m³.

A mix of 1:3:6 contains 10 parts:

- ◆ 1 part cement
- ◆ 3 parts fine aggregate
- ◆ 6 parts coarse aggregate.

To determine dry materials for a given quantity of concrete multiply volume by density, which gives total mass, and then divide by the number of parts. This gives the amount of cement; multiply cement by 3 to give fine aggregate and then multiply cement by 6 to give coarse aggregate.

For 2.5 m³ of 1:3:6 concrete:

Mass of concrete $= 2.5 \times 2400$
$= 6000$ kg

Number of part $= 10$

Amount of cement $= 6000 \div 10$
$= 600$ kg or 24×25 kg bags

Amount of fine aggregate   = 600 × 3

   = 1800 kg, say almost two 1 tonne bags or 72 × 25 kg bags

Amount of coarse aggregate   = 600 × 6

   = 3600 kg, say almost four 1 tonne bags or 144 × 25 kg bags

Tiles and paving slabs

For a particular area these can be calculated using the following method:

◆ Determine the total area to be covered in m² (excluding any openings)
◆ Determine the total area of a tile or slab in m²
◆ Divide the area to be covered by the area of the tile or slab
◆ Add percentage for cutting and waste (typically 5 to 10%).

To determine the number of 1500 mm square tiles required for a kitchen floor.

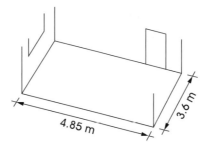

Figure 5.118

Plan of room

Area of floor   = 3.6 × 4.85

   = 17.46 m²

Area of tile   = 0.15 × 0.15

   = 0.0225 m²

Number of tiles   = 17.46 ÷ 0.0225

   = 776 tiles

7½% cutting and waste   = 776 × 1.075

   = 835.2, say 835 tiles

Paint

For a particular area paint required can be calculated using the following method.

◆ Determine the total area to be covered in m² (excluding any openings).
◆ Divide the total area by the recommended covering capacity (coverage) of the paint to be used. This will give the number of litres required for one coat.
◆ Multiply litres for one coat by number of coats, to give total litres required.

Determine the amount of paint required to paint the wall of a factory with two coats of emulsion. Typical manufacturer's coverage figures are:

◆ Primers and undercoats cover 12–14 m² per litre per coat
◆ Gloss or satin top coats cover 14–16 m² per litre per coat
◆ Emulsions cover 10–12 m² per litre per coat.

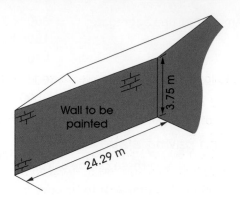

Figure 5.119

Area of wall to be painted

Use the minimum coverage from the table for a brick wall.

Area of wall	= 24.29 × 3.75
	= 91.0875 m²
Amount of paint for 1 coat	= 91.0875 ÷ 10 (coverage)
	= 9.10875 litres
Total amount of paint	= 9.10875 × 2 (coats)
	= 18.2175 litres, say 20 litres or 4 × 5 litre tins.

Costing materials

This can be carried out once the required quantities of material have been calculated. It is a simple matter of finding out prices and multiplying these by the number of items required.

Suppose you were asked to find the total price of four 2440 × 1220 sheets of 18 mm MDF. From the typical extract of the supplier's price list these are £10.97 each including VAT.

Total price	= Price per sheet × Number of sheets
	= £10.97 × 4
	= £43.88

Sheet Materials				*Price per item*	
Product	Size	Product code		£ inc. VAT	£ exc. VAT
Blockboard BB/CC 18 mm	2440 × 1220 mm	5022652518455		20.32	17.29
PLYWOOD					
Hardwood WBP BB/CC 4 mm	2440 × 1220 mm	5022652504588		7.00	5.96
Hardwood WBP BB/CC 6 mm	2440 × 1220 mm	5022652506605		8.50	7.23
Hardwood WBP BB/CC 9 mm	2440 × 1220 mm	5022652519629		11.55	9.83
OSB					
OSB2 11 mm	2440 × 1220 mm	5022652560386		7.41	6.31
OSB3 18 mm	2440 × 1220 mm	5014957148673		15.98	13.60
MDF					
MDF 6 mm	2440 × 1220 mm	5022652550035		6.57	5.60
MDF 12 mm	2440 × 1220 mm	5022652512231		8.48	7.22
MDF 18 mm	2440 × 1220 mm	5022652518257		10.97	9.34
HARDBOARD					
Standard Hardboard 3.2 mm	2440 × 1220 mm	5022652503314		2.60	2.21
White Faced Hardboard 3.2 mm	2440 × 1220 mm	5022652503338		5.17	4.40
CHIPBOARD					
Flooring Grade 18 mm P4	2400 × 600 mm	5014957105706		3.98	3.39
Flooring Grade 22 mm P4	2400 × 600 mm	5014957105720		5.58	4.75
Flooring Grade 18 mm P5	2400 × 600 mm MR	5014957054691		5.31	4.52
Flooring Grade 22 mm P5	2400 × 600 mm MR	5014957088320		7.98	6.79
Standard Grade 12 mm	2440 × 1220 mm	5014957054677		4.95	4.21

Figure 5.120

Sheet material price list

The following materials are required for a small building extension:

◆ 12 off 2.7 m long 50 × 200 joists
◆ 8 off 2.7 m long 50 × 100 sawn softwood
◆ 12 off 50 mm joist hangers
◆ 11 off 600 × 2440 T&G chipboard flooring
◆ 1750 off Redland red drag faced bricks
◆ 450 off 100 mm blocks
◆ 31 bags of cement
◆ 3 tonne of fine aggregate (building sand)
◆ 2 tonne of fine aggregate (sharp sand)
◆ 4 tonne of coarse aggregate (gravel 20 mm).

On phoning your material supplier, the prices shown on the telephone message pad were obtained.

Use this information to determine the following:

a) Total cost at list price
b) Trade discount of 7½% on total cost
c) VAT at 17½% on discounted total cost
d) Actual cost payable including VAT.

Telephone Message

Date **26 OCT**   Time **9.45**

Message for **JAMES**

Message from (Name) **BBS SUPPLIES**

(Address) **LAKESIDE INDUSTRIAL PARK,**

NOTTINGHAM

(Telephone) **01159 464368**

Message **All plus VAT at 17.5%**

50 × 200 Joists	£2.35 per m
50 × 100 Sawn	£0.85 per m
Joist Hangers	£1.69 each
600 × 2400 Chipboard Flooring	£3.98 each
Redland Brick	£0.70 each
100 mm Blocks	£1.62 each
Cement	£4.99 per bag
Building Sand	£1.06 per 25 kg
Sharp Sand	£1.14 per 25 kg
Gravel	£24 per tonne

Message taken by **CHRIS**

Figure 5.121

44. Calculate the number of eight-hour days required for a four-person gang of bricklayers to build 145 m² of brickwork. (Use the rate of 0.3 m² per bricklayer per hour.)

45. A = B × C; find the value of B if A = 4.5 m² and C = 1.2 m.

46. The following materials are required for a refurbishing contract:

Softwood
Sawn softwood at £158.50 per m³

Item	Number	Size
Joists	16	50 × 225 × 3600
Strutting	10	50 × 50 × 4200
Studwork	84	50 × 100 × 2400
Battening	50	50 × 50 × 4800

Flooring
18 mm flooring grade chipboard at £49 per 10 m²; 30 sheets 600 × 2400 mm

Calculate the total cost including an allowance of 10 per cent for cutting and wastage and 17½ per cent for VAT.

47. What radius is required to set out a centre for a segmental arch having a rise of 550 mm and a span of 4.500 m?

48. What is the diameter of a circular rostrum if its perimeter measures 12 m?

49. A rectangular room 4.200 m × 6 m is to be floored using hardwood boarding costing £9.55 per square metre.
a) Allowing 12½ per cent for cutting wastage, how many square metres are required?
b) What would be the total cost including 17½ per cent VAT?

50. A builders' merchant sells cement in three sizes. Which is the best value?

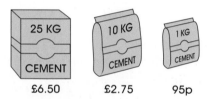

Figure 5.122

25 KG CEMENT £6.50   10 KG CEMENT £2.75   1 KG CEMENT 95p

51. A window is in the shape of a rectangle and a semicircle. Find the area of the window.

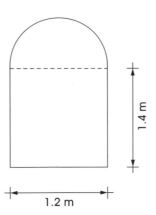

Figure 5.123

1.4 m

1.2 m

52. Find the angles, and give the reason:

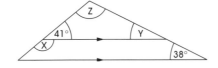

Figure 5.124

41° Z Y X 38°

 a) Angle X
 b) Angle Y
 c) Angle Z

53. Details of a one-brick thick garden wall 15.750 m long are illustrated.

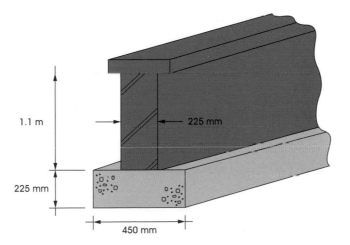

1.1 m

225 mm

225 mm

450 mm

Figure 5.125

Calculate:
 a) The number of bricks required to build the wall
 b) The amount of cement and fine aggregate required to build the wall using a 1:6 mix
 c) The total volume of concrete for the foundation
 d) The number of coping stones if each covers a length of 600 mm.

54. Use the following time sheet to determine how much Ivor Carpenter has earned for a week. Take the rate of pay for weekdays as £9.25 for the first 8 hours each day and time and a half after that. Saturdays is also paid at time and a half and Sundays at double time (enter your results on a photocopy).

BBS Recruitment Solutions: WEEKLY TIME SHEET		Name: I. CARPENTER			Works No. 26	
Day	Date	Start Time	Lunch	Finish Time	Total basic hours	Total overtime hours
Monday	15/3	7.30	½ hr	6.00		
Tuesday	16/3	7.30	½ hr	5.30		
Wednesday	17/3	8.00	½ hr	6.15		
Thursday	18/3	8.00	½ hr	6.00		
Friday	19/3	7.30	½ hr	5.45		
Saturday	20/3	6.00		12.00		
Sunday	21/3	6.00		12.00		
Signature:	I Carpenter					

Figure 5.126

55. What would be the net pay in Question 54 if a total of 26% deductions were taken from the gross pay?

56. Determine for the room shown:
   a) the number of 900 × 1800 mm sheets of plasterboard required to cover the ceiling
   b) the total length of skirting required allowing 10% for cutting and waste
   c) the total length of timber required to cover the floor if the boards have a covering width of 95 mm
   d) the amount of paint required for two coats of emulsion to the walls and ceilings if 1 litre covers 10 m², making a reduction for the door and the window.

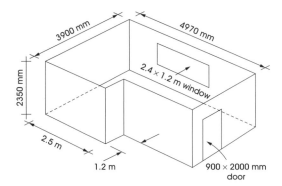

3900 mm
4970 mm
2350 mm
2.4 × 1.2 m window
2.5 m
1.2 m
900 × 2000 mm door

Figure 5.127

57. A cavity wall consists of a half-brick thick outer skin and a 100-mm thick blockwork inner skin. Calculate the number of bricks, blocks and mortar required for the cavity wall, 9 m long and 4.8 m high.

58. A 1:6 mortar mix is used for Question 44. Determine the amount of cement and fine aggregate required.

59. A wall area 3.6 m × 2.4 m containing a window 1.2 m × 900 mm is to be tiled using 100-mm square tiles. Calculate the number of tiles required including an allowance of 5% for cutting and waste.

Index